SOME METHODS FOR MICROBIOLOGICAL ASSAY

THE SOCIETY FOR APPLIED BACTERIOLOGY
TECHNICAL SERIES NO. 8

SOME METHODS FOR
MICROBIOLOGICAL ASSAY

Edited by
R. G. BOARD
*School of Biological Sciences, University of Bath,
Claverton Down, Bath, England*

AND

D. W. LOVELOCK
*H. J. Heinz Co. Ltd., Hayes Park, Hayes, Middlesex,
England*

1975

ACADEMIC PRESS
LONDON   NEW YORK   SAN FRANCISCO
A Subsidiary of Harcourt Brace Jovanovich, Publishers

ACADEMIC PRESS INC. (LONDON) LTD
24-28 OVAL ROAD
LONDON, NW1

U.S. Edition published by
ACADEMIC PRESS INC.
111 FIFTH AVENUE,
NEW YORK, NEW YORK 10003

Library of Congress Catalog Card No : 74–18518
ISBN : 0–12–108240–7

Printed in Great Britain by
Cox & Wyman Ltd., Fakenham, Norfolk, England

Contributors

J. ASHWORTH, *B.F.M.I.R.A., Randalls Road, Leatherhead, Surrey, England*

A. C. BAIRD-PARKER, *Unilever Research Laboratory, Colworth House, Sharnbrook, Bedford, England*

SHOSHANA BASCOMB, *Department of Biochemistry, Imperial College of Science and Technology, London SW7 2AZ, England*

H. A. BEHAGEL, *Microbiological Research and Development Laboratories, Organon, Scientific Development Group, Oss, The Netherlands*

TH. M. BERG, *Microbiological Research and Development Laboratories, Organon, Scientific Development Group, Oss, The Netherlands*

M. E. BLACK, *Biochemistry Department, A.R.C. Institute of Animal Physiology, Babraham, Cambridge, England*

I. J. BOUSFIELD, *National Collection of Industrial Bacteria, Torry Research Station, Ministry of Agriculture, Fisheries and Food, Aberdeen, Scotland*

A. F. BRAVERY, *Building Research Establishment, Princes Risborough Laboratory, Princes Risborough, Aylesbury, Bucks, England*

D. R. CLIFFORD, *University of Bristol, Long Ashton Research Station, Long Ashton, Bristol BS18 9AF, England*

J. M. DEN BURGER, *Microbiological Research and Development Laboratories, Organon, Scientific Development Group, Oss, The Netherlands*

C. J. EVANS, *Pharmaceutical Production Control Laboratory, Beecham Research Laboratories, Worthing, Sussex, England*

G. G. FOWLER, *Alpin & Barret Ltd., Yeovil, Somerset, England*

C. A. GRANTHAM,* *Department of Biochemistry, Imperial College of Science and Technology, London SW7 2AZ, England*

E. GROSSBARD, *ARC Weed Research Organization, Begbroke Hill, Yarnton, Oxford OX5 1PF, England*

P. HAMILTON, *Beecham Research Laboratories, Walton Oaks, Dorking Road, Tadworth, Surrey, England*

LINDA L. HARGREAVES, *B.F.M.I.R.A., Randalls Road, Leatherhead, Surrey, England*

E. C. HISLOP, *University of Bristol, Long Ashton Research Station, Long Ashton, Bristol BS18 9AF, England*

R. HOLBROOK, *Unilever Research Laboratory, Colworth House, Sharnbrook, Bedford, England*

B. JARVIS,* *B.F.M.I.R.A., Randalls Road, Leatherhead, Surrey, England*

H. H. JONES,** *Thames Water Authority, New River Head Laboratories, 177 Rosebery Avenue, London EC1R 4TP, England*

P. NOONE,*** *Department of Bacteriology, School of Pathology, Middesex Hospital Medical School, Riding House Street, London W1P 7LD, England*

G. H. PALMER, *Beecham Research Laboratories, Walton Oaks, Dorking Road, Tadworth, Surrey, England*

J. R. PATTISON, *Department of Bacteriology, School of Pathology, Middlesex Hospital Medical School, Riding House Street, London W1P 7LD, England*

L. B. PERRY, *National Collection of Industrial Bacteria, Torry Research Station, Ministry of Agriculture, Fisheries and Food, Aberdeen, Scotland*

S. F. B. POYNTER, *Thames Water Authority, New River Head Laboratories, 177 Rosebery Avenue, London EC1R 4TP, England*

ANTHEA ROSSER, *B.F.M.I.R.A., Randalls Road, Leatherhead, Surrey, England*

J. M. SHEWAN, *National Collection of Industrial Bacteria, Torry Research Station, Ministry of Agriculture, Fisheries and Food, Aberdeen, Scotland*

R. C. B. SLACK, *Department of Bacteriology, School of Pathology, Middlesex Hospital Medical School, Riding House Street, London W1P 7LD, England*

J. S. SLADE, *Thames Water Authority, New River Head Laboratories, 177 Rosebery Avenue, London EC1R 4TP, England*

W. ROBERT SPRINGLE, *Paint Research Association, Waldegrave Road, Teddington, Middlesex TW11 8LD, England*

D. A. SYKES, *Pharmaceutical Production Control Laboratory, Beecham Research Laboratories, Worthing, Sussex, England*

J. TRAMER, *Unigate Ltd., Central Laboratory, Western Avenue, London W3, England*

R. W. WHITE, *Biochemistry Department, A.R.C. Institute of Animal Physiology, Babraham, Cambridge, England*

G. I. WINGFIELD, *ARC Weed Research Organization, Begbroke Hill, Yarnton, Oxford OX5 1PF, England*

S. J. L. WRIGHT, *School of Biological Sciences, University of Bath, Bath BA2 7AY, Avon, England*

* Present address: Research and Development Division, G. D. Searle & Co Ltd., Lane End Road, High Wycombe, Bucks HP12 4HL, England.
** Present address: The Department of Biological Sciences, University of Surrey, Guildford, England.
*** Present address: Bacteriology Department, Royal Free Hospital, London WC1, England.

Preface

THIS volume includes contributions to the Autumn Demonstration Meeting of the Society for Applied Bacteriology held on October, 1962, at the Unigate Central Laboratories, Western Avenue, London, W3. It is Number 8 in the Technical Series and continues the Society's policy of encouraging members and guests to exhibit methods which are of great use in the day-to-day work of a laboratory. The demonstrators have described their methods in this book which is intended to be a reference source at the bench. We wish to thank them all for the great effort which they took both in the preparation of their demonstrations and for their contributions in this book.

Our thanks go also to The Chief Scientist of Unigate Central Laboratories, Dr. D. B. Gammack, and Miss W. Hyett-Brown and Dr. W. A. Cox for all their help with the laboratory arrangements.

November 1974

R. G. BOARD
D. W. LOVELOCK

Contents

Assay of Myo-inositol using the Yeast *Kloeckera apiculata* (*K. brevis*)

R. W. WHITE AND M. E. BLACK

Notes on the Preservation and Checking of Vitamin Assay Bacteria

L. B. PERRY, I. J. BOUSEFIELD AND J. M. SHEWAN

A Simple, Rapid Assay for the Measurement of Antibiotic Concentrations in Human Serum

P. NOONE, J. R. PATTISON AND R. C. B. SLACK

Bio-assay of Agricultural and Horticultural Fungicides

E. C. HISLOP AND D. R. CLIFFORD

Use of Micro-Algae for the Assay of Herbicides
S. J. L. WRIGHT

Semi-automated Method for Microbiological Vitamin and Antibiotic Assays

TH. M. BERG, J. M. DEN BURGER, H. A. BEHAGEL

*Microbiological Research and Development Laboratories,
Organon, Scientific Development Group,
Oss, The Netherlands*

Microbiological turbidimetric assays of vitamins and antibiotics consist of mainly tedious and laborious procedures that may easily lead to the introduction of small mistakes which result finally in assay failure. Standardization of all manipulations and procedures involved is the only solution for this problem. Because automation can provide an important contribution to this goal, many attempts have been made to design automated systems for microbiological turbidimetric assays.

Several continuous flow systems (Gerke *et al.*, 1960; Haney *et al.*, 1962; Shaw and Duncombe, 1963; Dewart, Naudts and Lhoest, 1965; Platt, Gentile and George, 1965) and discrete sample systems (McMahan, 1965; Gualandi and Morisi, 1967; Kuzel and Kavanagh, 1971; Berg and Behagel, 1972) have been described. The first type applied mainly for antibiotic assays, the latter also for the assay of vitamins. In the continuous flow systems the rather short incubation period needed for antibiotic assays is included in the system. In most discrete sample systems, preparation of the sample and the final turbidity reading are automated separately. The incubation period, up to 18 h for vitamins, is excluded from automation. Recently the first commercial system (Autoturb; Kuzel and Kavanagh, 1971), based solely on the latter principle, has become available. The diluter module of the Autoturb prepares automatically up to 4 test mixtures from a pre-diluted sample.

The presently described system prepares the test mixtures by the continuous flow method. In this way a large part of the sample preparation including a 1000-fold dilution step is automated. The system described has been used for routine microbiological turbidimetric assays for both vitamins and antibiotics for more than 5 years. It consists of two separate units, one for sample preparation and one for turbidity measurement.

Instrumentation

Dilution and dosage unit

Principle

The system (Fig. 1) consists of a main sampler (26 × 36 × 30 cm), 4 auxiliary samplers, 2 tubing proportioning pumps with a capacity for 12 tubes each, manifold, and combicollector. All units are electrically linked to a control unit (45·5 × 23·5 × 14·5 cm).

FIG. 1. Dilution and dosage unit consisting of main sampler, 4 auxiliary samplers, 2 proportioning pumps, manifold control unit and combicollector.

The system differs fundamentally from the autoanalyser system in that the final step consists of the collection of the mixture prepared in previous steps instead of a measurement of some kind. For this purpose the dilution and dosage unit has been provided with: (a) 4 auxiliary samplers supplying the fluids needed for sample preparation; (b) a combicollector to collect the prepared samples in tubes, and (c) a control cabinet to programme the action of samplers and combicollector (two time clocks for each action).

The main and auxiliary samplers are programmed to draw sample, diluent, and test broth liquid respectively for equal periods, thus forming liquid segments of equal length. By adjustment of the time clocks, synchronization of streams can be attained so that all liquid parts meet precisely at the intersections in the manifold. In the first part of the manifold each sample segment is diluted 1000-fold. The diluted sample segment is divided into two parts. One part is once more diluted and then both parts are mixed with test broth thus forming a low and high potency dose respectively. These segments arrive simultaneously at the end of the manifold where the valves of the combicollector are programmed to fill the central part of the liquid segments into the tubes. The unit has a capacity for diluting 40 samples thus producing 80 test mixtures in tubes per hour.

Main sampler

The main sampler (Fig. 2) consists of a rack holder, for racks with sample tubes, a carriage carrying the sampling needle, and a cup for rinsing fluid. The carriage moves automatically from one sample tube (and row) to

FIG. 2. Main sampler with handles for positioning the sample needle manually. The sample carriage provided with the cup for rinsing liquid. At the top the adjustment screw for the height of the needle.

another. With two small handles in the side panel of the carriage, the sampling needle can also be positioned manually. If necessary, the last rack in the rackholder can be replaced by a reservoir with rinsing fluid. The capacity is 6 racks (3 × 20 × 6·5 cm) of 20 tubes (volume, 3 ml). The sampling needle can be adjusted with a screw to the required height. Sampling and rinsing time can be set by means of time clocks mounted in a separate module in the control unit.

Auxiliary samplers

Four auxiliary samplers (Fig. 3) serve for the supply of diluent (H1, H2, H4) and test broth (H3). In contrast to the main sampler, where rinsing fluid is drawn between the samples, the auxiliary samplers draw air when they are not sampling fluid. Air is drawn between sampling to enable

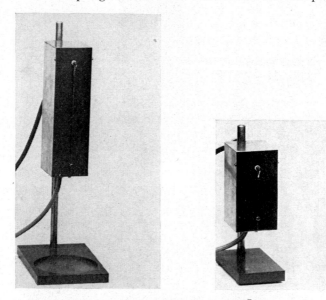

F IG. 3. Auxiliary samplers on the right for sampling out of a small reservoir; on the left for sampling out of a large vessel.

an easier check on the flow of the liquid segments and to economize on the use of the expensive test broth. The construction of the auxiliary samplers is similar to the sampling device mounted on the main sampler. Two types of auxiliary samplers exist. One type is used for sampling diluent out of a small reservoir. The difference between highest and lowest position of the needle being 4·5 cm. Three samplers of this type are fixed to a common support. The second type is a larger sampler for sampling out of

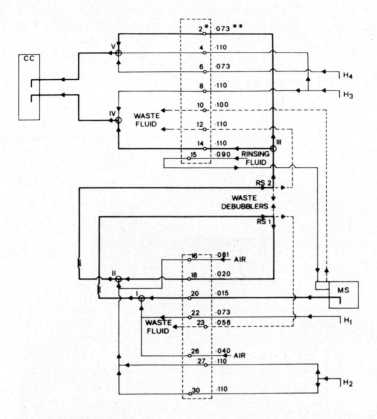

FIG. 4. Scheme of Dilution and dosage unit. (*) Position on proportioning pump. (**) Internal diameter of tubes. MS = main sampler provided with cup for rinsing fluid; H1 = auxiliary sampler diluent first step; H2 = auxiliary sampler diluent second step; H3 = auxiliary sampler test broth; H4 = auxiliary sampler diluent low potency samples; CC = combicollector. Flow of sample stream is indicated by a thick line.

Intersection I:	dilution 20 fold	
RS 1:	re-sampling	1000 fold
Intersection II:	dilution 50 fold	
RS 2:	resampling	

Intersection III: sample divided into two parts, so that 4 ml and 2 ml from the originally 1000-fold diluted sample arrives at intersections IV and V respectively.

Intersection IV: addition of 4 ml test broth (high potency).

Intersection V: addition of 2 ml diluent and 4 ml test broth (low potency).

In the combicollector the central part (4 ml) of high and low potency samples are simultaneously dispensed in two separate tubes.

a bigger vessel. The difference between highest and lowest position of the needle is 15 cm. The action of each auxiliary sampler is controlled by time clocks. These are assembled for each auxiliary sampler in a separate module in the control cabinet.

Proportioning pump and manifold

The proportioning pumps (Cenco N.V. The Netherlands) each have a capacity of 12 tygon tubes. The manifold (Figs 4 and 5) has been built up from tygon tubing with diameters up to 2·8 mm and glass connections. For practical reasons the manifold has been divided in two parts. One part

FIG. 5. Two proportioning pumps and manifold built up partly outside and partly inside the Perspex box. Note the small glass connectors at the outlet of the pumps and those fixed in the box.

consists of the tubing on the proportioning pumps and some attached glass connections for junctions of liquid and air streams. The other part of the manifold has been built in a Perspex box (19 × 11 × 9 cm), containing the remainder of the tubing. The glass connections are fixed in positions to provide for optimal flow. Both parts of the manifold are constructed separately; they can be easily disconnected and replaced. All waste tubes are connected to a larger tube fixed on the outside of the box, which is connected to the waste. The special construction of the proportioning pumps (Cenco?) with the in- and outlets of the tubings positioned at the same side enables the construction of a manifold with a minimal length of the tubing. For this purpose special small glass connections were constructed. It takes 4·5 min for a sample to run through the tubing circuit. The two pumps and the Perspex box are fixed on a base plate (30 × 45 cm).

The combicollector

The combicollector (Fig. 6) serves to dispense the final reaction mixtures into the tubes. It consists of a conveyor belt above which are mounted two solenoid valves each connected to the bypass of a stainless-steel outlet. In the "on"-position the valves are open, and thus connect the

FIG. 6. Combicollector. Conveyor belt with slide provided with two valves with outlets at one tube interval. Switches serve to operate separately the combicollector and the alarm system.

bypass to the waste. When the valves are closed the liquid passes to the outlets. Special stainless-steel racks (1·6 × 15 cm) each containing two rows of six culture tubes are carried by the conveyor belt in such a way that two tubes are centred below the outlets of the valves. Each time the valves are closed the liquid is dispensed into two tubes. As the valves are opened the next two culture tubes are moved to this position. The conveyor belt can hold 4 racks at a time and is provided with a slide. The valves and the transport system are controlled by two time clocks of the control cabinet. The combicollector is provided with an alarm which gives a signal when no racks are on the conveyor belt under the outlets. The alarm as well as the actions of the combicollector can be switched off by two switches on the combicollector.

Control cabinet

The control cabinet (Fig. 7) consists of two standard frames each containing 4 plug-in modules built up with Transistor-Transistor Logic (TTL)

FIG. 7. Control cabinet; two frames containing 8 plug-in modules (numbered from top left to bottom right). The pair of time-clocks in the third module serves for controlling the rinsing time, 30 sec fixed, and sampling time, 60 sec fixed. Time-clocks in modules 4, 5, 6 and 8 control auxiliary samplers H1–H4. (Time-clocks delay time: 0 to 30 sec adjustable; time-clocks for sampling operation 60 sec fixed.) The lower pair of time-clocks in the 8th module controls the extra rinse cycle performed by H2. Delay time adjustable from 0–50 sec, sampling time 10 sec fixed. The time-clocks in the 7th module control the combicollector. Delay time 0–60 sec adjustable. Operation of time-clocks delay time can be derived from start or end of the sampling time of main sampler with the switches on modules 3–8.

integrated circuits. The modules are interchangeable. The logic power and main power are housed in two of them. The remaining 6 modules serve for controlling the operations of the individual elements, e.g. main sampler, 4 auxiliary samplers and the combicollector. Control is by electronic time-clocks which are adjustable by means of thumbwheels.

The sampling time and the rinsing time of the main sampler are controlled by the time-clocks of one module. Four pairs of time-clocks, each pair in one individual module, serve for controlling the three auxiliary samplers, H1, H3, and H4, and the opening of the valves on the combicollector. The remaining 4 time-clocks control the two operations to be performed by auxiliary sampler H2 (see procedure).

The first time-clock in each pair controls the delay time, the second one controls the actual operating period. The period between the receipt of the electrical signal and the start of the operation is defined as the delay time. Signals derived from the time-clocks of the main sampler initiate the start of all delay times with the exception of the time-clocks controlling the delay time for the second operation of auxiliary sampler H2. In the first case, the starting signal can be derived from the start or end of the sampling time of the main sampler. In the case of auxiliary sampler, H2, the signal is derived from the end of the first operation of itself. The time-clocks controlling the duration of the actual operations have been fixed on 60 sec for most sampling actions (H1, H3, H4 and main sampler), on 35 sec for the combicollector, on 4 and 10 sec respectively for the delay and second sampling of auxiliary sampler H2. The delay time for the other individual operations can be varied. In this way means are provided for synchronization of the various liquid streams in the manifold. In routine use, only minor adjustments of the delay times are necessary to ensure accurate synchronization.

Turbidity measurement unit

Principle

The unit (Fig. 8) consists of: a transport system connected with a sample transfer unit (1·25 × 50 × 50 cm) a turbidimeter, a digilog convertor and a printer.

The unit measures samples at a rate of 300 tubes/h. The transport system

FIG. 8. Turbidity measurement unit consisting of sampling and stirring devices built on to a conveyor belt, a turbidimeter, a digilog convertor and a printer.

has been designed similarly to that used in the combicollector. The culture tubes can thus be handled in the same racks throughout the assay. The content of the tubes is homogenized prior to measurement so that the cultures can be measured automatically without any additional individual stirring of the tubes.

Transport system with sample transfer unit

The conveyor belt of the Turbidity measurement unit has been similarly constructed to that of the combicollector. The stirrer (Philips N.V., The Netherlands) and the sample needle are mounted together on a bar in the

FIG. 9. The stirrer and sample needle mounted on a conveyor belt. Racks move from right to left.

middle of the conveyor belt (Fig. 9). The stirrer and the sample needle are introduced simultaneously into the respective culture tubes, the stirrer in a position 6 tubes before the tube where the sampling is performed. The motor of the stirrer is programmed to switch on for two periods of about 2 sec. Homogenization of the contents takes place while the stirrer has reached its lowest point in the tube. In a position half-way up the height of the tube the stirrer once again functions to throw off all adhering liquid so reducing any carry-over to the next tube.

A modified Automatic Sample Transfer apparatus (AST 100, Vitatron, The Netherlands) is built over the transport system drawing the samples through the flow-cell in the Vitatron colorimeter. The flow-cell (type 186, Hellma GmbH, Germany) has a light path of 1 cm and a volume of 0·5 ml. For turbidity reading a wavelength of 853 nm is used. The first part of the sample, c. 1 ml, is used for rinsing the flow-cell, the second part, c. 0·5 ml, is used for turbidity measurement. In total, c. 1·5 ml of liquid is used for each measurement. The disadvantage of more static systems in which flow birefringence may cause inconsistent readings is not encountered here because of the high flow rate. The transmission values are converted into extinction values by a digilog convertor and are printed on an Addo-X calculator.

Test procedure

The reference standard and the samples of substances to be determined are diluted to the approximate level of 2000 times the concentration to be obtained in the culture tube (high dose level). Tubes filled with these concentrated solutions are placed in the racks of the main sampler so that, after the last sample, the carriage of the main sampler arrives at the rinsing fluid reservoir.

The main sampler draws sample and rinsing fluid (double distilled water containing 0·01% (v/v) Triton X 405) during 60 and 30 sec respectively. The auxiliary samplers also sample liquid for 60 sec thus providing fluid segments of equal length. By adjusting the delay times for the individual samplers, the continuous-flow streams can be synchronized so that all liquid parts meet precisely at the various intersections in the manifold (Figs 4 and 5). At intersections I and II, the samples are diluted 20- and 50-fold with diluent from auxiliary samplers H1 and H2 respectively. After the second dilution step, the diluted sample is divided into two parts at intersection III in the manifold. One part (4 ml) goes to intersection IV and is mixed with an equal volume of test broth from H3 to form the high dose level. The second part (2 ml) is mixed at intersection V with 2 ml of diluent (from H4) and with 4 ml test broth (from H3) thus forming the low dose level.

Both sample streams are then dispensed simultaneously into culture tubes by the combicollector as follows. The two ends of the tubing circuit are connected with the valves on the combicollector. These valves are controlled in such a way that from each liquid segment of *c.* 8 ml the first 3 ml are discarded as waste. The valves are then closed for 35 sec and *c.* 4 ml from the central part of both liquid segments are dispensed into the culture tubes. The remainder of the liquid is led to waste. Only the central part of the entire segment is used in order to avoid contamination from the preceding samples and slight differences in composition due to minor fluctuations in synchronization.

Rinsing fluid sampled by the main sampler serves to avoid contamination of a sample segment with the preceding one in the sample line up to intersection II. From that point on, the manifold is rinsed with dilution fluid from auxiliary sampler H2. For this purpose the latter sampler is programmed to sample dilution liquid during a period of 10 sec between its main sampling periods. To limit carry-over between two consecutive samples, care was taken to build a circuit with minimal total length.

After each assay run the manifold is rinsed with 5% (v/v) aqueous formaldehyde, 0·2% (v/v) aqueous sodium lauryl sulphate and finally with dilution fluid. Vitamin and antibiotic assays can be performed alternately on the same manifold without additional precautions. Every fortnight an additional rinse with 2N H_2SO_4 is carried out. Every 2 months the tubes of the manifold are renewed.

After each row of culture tubes has been filled the 6 tubes are covered with a stainless steel cover. To avoid premature growth in the culture tubes, the racks are placed in cold water (0°) immediately after filling. The racks with tubes filled with the vitamin/antibiotic cultures on the combicollector are transported in special trays and incubated either in a thermostatically controlled water bath (Marius N.V., The Netherlands) or on a shaker (type SL 69, Marius N.V., The Netherlands) in a thermostatically controlled cabinet. In the latter case, the racks are tilted on the shaker to increase aeration and agitation of the cultures.

After the incubation (3–5 h for antibiotic; *c.* 18 h for vitamin assays) growth is terminated by immersion of the racks in cold water (0°).

On the Turbidity measurement unit, the cultures are mechanically homogenized and the turbidity is read at a rate of 300 tubes/h. On both conveyor belts of the combicollector and the turbidity measurement unit, the tubes are handled in a similar sequence. From each of the racks containing 12 tubes, the front row of 6 tubes is filled, or sampled and then all the racks are turned and the other row of tubes is handled. Sample sequence numbers together with extinction values are printed on paper tape.

CYANOCOBALAMIN

MICROBIOLOGICAL TWO POINT PARALLEL LINE ASSAY

BASIC DATA STANDARD

SUBSTANCE : CYANOCOBALAMIN
DATE : 14- 3-73
ASSAY NUMBER : 25
NUMBER STANDARDSETS : 7
NUMBER UNKNOWNS : 15
CONCENTRATION RATIO : 2.00
*E. QUANTITY MG : 240.00
DILUTION(h) ML : 6.00 E6
DIFF.MAX : 155.

ACTUAL NUMBER OF STANDARDSETS : 7
ACTUAL NUMBER OF UNKNOWNS : 18

STANDARDSETS

NUMBER	1	2	3	4	5	6	7
CONSEC.NR	4	9	10	13	15	19	21
L1:	463	451	450	461	438	443	447
L2:	443	440	462	439	456	478	460
TURBIDITY READINGS							
M1:	681	662	637	657	647	678	680
M2:	672	648	678	682	662	649	641
MEAN	564.75	556.25	556.75	559.75	554.75	560.00	557.00
REGR.COEFF	742.45	695.94	669.37	729.16	689.30	661.06	687.64

VARIANCE(REGR.COEFF) : 883.4583
LIMITS REGR.COEFF : 618.67 AND 774.17
REGR.COEFF. S,U : 591.74
LAMBDA : 0.015

MEAN STANDARD : 557.04
REGR.COEFF. S : 696.42
TURBIDITY BLANKS BEFORE : 51 51 57 49 48
TURBIDITY BLANKS AFTER : 63 57 57 51 53

UNKNOWN PREPARATIONS OF ASSAY NUMBER : 25   CYANOCOBALAMIN

NUMBER	1	2	3	4	5	6	7	8	9
CONSEC.NUMBER	2	5	1	11	14	7	18	8	28
PREPAR-ATION	STAND A5	PREPARA-TION A	PREPARA-TION A	PREPARA-TION A	PREPARA-TION A	PREPARA-TION B	PREPARA-TION B	PREPARA-TION B	PREPARA-TION B
BATCH-NO	UNKNOWN	SOLUTION	SOLUTION	SOLUTION	SOLUTION	SOLUTION	SOLUTION	SOLUTION	SOLUTION
QUANTITY	1.00 MCG	1.00 ML	1.00 ML	1.00 ML	1.00 ML	1.00 ML	1.00 ML	1.00 ML	1.00 ML
ASS POT IN MCGS/QUANTITY	1.000	0.120	1.200	6.000	12.000	1.200	2.000	3.000	4.000
WE.QUANTITY	246.??	1.00	1.00	1.00	1.00	1.00	1.00	1.00	1.00
DILUTION(W)ML	5.00 F6	3.00 E3	3.00 F4	1.50 E5	3.00 F5	3.50 E4	5.00 E4	7.50 E4	1.20 E5
TURBIDITY READINGS L1	449	887	687	236	71	673	483	398	290
L2	457	872	673	247	87	681	521	417	301
H1	679	880	854	401	180	872	710	612	472
H2	652	871	852	412	153	869	699	671	478
MEAN	557.00	877.00	766.25	324.00	122.75	773.75	508.25	597.00	384.75
REGR.COEFF	698.96	-9.97	596.29	548.12	294.67	642.79	735.91	661.06	592.96
TEST ON // M	-0.000	0.463	0.342	-0.337	-0.628	0.313	0.069	-0.072	-0.249
L/2	0.000	**********	**********	**********	**********	0.021	0.022	0.020	**********
Q	0.9999	2.9910	2.4465	0.4604	0.2356	2.0573	1.1470	0.8466	0.5636
X (MCGS/QUANT)	1.0000	0.1200	1.2000	6.0000	12.0000	1.2000	2.0000	3.0000	4.0000
0.95 LIMITS LOW	0.96	**********	**********	**********	**********	2.35	2.19	2.43	**********
HIGH	1.05	**********	**********	**********	**********	2.59	2.44	2.66	**********
POTENCY	1.00	0.35	2.4	2.0	2.8	2.47	2.29	2.54	2.7
DIMENS.= MCGS/QUANTITY	MCG/ 1.00 MCG	MCG/ 1.00 ML	MCG/ 1.00 ML	MCG/ 1.00 ML	MCG/ 1.00 ML	MCG/ 1.00 ML	MCG/ 1.00 ML	MCG/ 1.00 ML	MCG/ 1.00 ML

CYANOCOBALAMIN

UNKNOWN PREPARATIONS OF ASSAY NUMBER : 25

NUMBER	10	11	12	13	14	15
CONSEC. NUMBER	12	22	16	17	3	6
PREPARATION	PREPARATION C	PREPARATION C	PREPARATION C	PREPARATION D	PREPARATION D	PREPARATION D
BATCH NO	COAT TBL	COAT TBL	COAT TBL	COAT TBL	COAT TBL	COAT TBL
QUANTITY	1.00 CTB	1.00 CTB	1.00 CTB	1.00 CTB	1.00 CTB	1.00 CTB
ASS POT IN MCGS/QUANTITY	2.000	2.000	2.000	10.000	10.000	10.000
WE.QUANTITY	1.00	1.00	1.00	1.00	1.00	1.00
DILUTION(M)ML	5.00 E4	5.00 E4	5.00 E4	2.50 E5	2.50 E5	2.50 E5
TURBIDITY READINGS L1	458	468	452	442	461	455
L2	461	457	467	484	449	467
H1	680	663	672	673	658	677
H2	672	679	661	663	672	666
MEAN	567.75	566.75	563.00	564.50	558.00	566.25
REGR.COEFF	719.26	692.62	587.64	687.64	697.60	699.27
TEST ON //						
X	0.015	0.014	0.009	0.011	0.014	0.013
L/2	0.020	0.020	0.020	0.020	0.020	0.020
K	1.0363	1.0329	1.0201	1.0252	1.0099	1.0311
K (MCGS/QUANT)	2.0000	2.0000	2.0000	10.0000	10.0000	10.0000
0.95 LIMITS LOW	1.98	1.97	1.95	9.8	9.7	9.9
HIGH	2.17	2.16	2.13	10.7	10.6	10.8
POTENCY	2.07	2.07	2.04	10.3	10.1	10.3
DIMENS.= MCGS/QUANTITY	MCG/ 1.00 CTB	MCG/ 1.00 CTB	MCG/ 1.00 CTB	MCG/ 1.00 CTB	MCG/ 1.00 CTB	MCG/ 1.00 CTB

25 PAGE 4

```
PREPARA- TION C                    PREPARA   TION D
COAT TBL                           COAT TBL
CYANOCOBALAMIN                     CYANOCOBALAMIN
ASSAY NO  25                       ASSAY NO  25

L/2 :    0.020                     L/2 :    0.020
1   :   2.1   (  2.0  -  2.2  )    1   :   10.2  (  9.8  -  10.7 )
2   :   2.1   (  2.0  -  2.2  )    2   :   10.1  (  9.6  -  10.6 )
3   :   2.0   (  1.9  -  2.1  )    3   :   10.3  (  9.9  -  10.8 )


    MEAN:  2.1    MCG/ 1.00 CTB        MEAN:  10.2   MCG/ 1.00 CTB
```

Ø = ZERO

FIG. 10. Standard computer print-out of a cyanocobalamin assay.
Page 13 (page 1 on print-out).
Basic data of assay and standardsets.
An indication of the precision of the assay is given by lambda $= \frac{s}{b}$, where s is the standard deviation of the mean extinction values of the standardsets, and b is the mean regression coefficient of the standardsets, and those unknowns that have not been rejected.
An unknown is rejected when the regression coefficient falls without the limits regression coefficient calculated for the pairs of standardsets. Accordingly test on parallelity is negative (indicated with *) 1/2 L and 0·95 confidence limits are not calculated.
Pages 14 and 15 (pages 2 and 3 on print-out).
Basic data and results of preparations.
Preparation A: The potency of cyanocobalamin is totally unknown. To obtain an indication about the potency 4 different dilutions have been assayed (3,000, 30,000, 150,000 and 300,000 fold, nr. 2, 3, 4 and 5 respectively). The turbidity of the low concentration of the 30,000 fold diluted sample (nr. 3) approximates the high concentration of the standard sets. The high concentration of the 150,000 fold diluted sample (nr. 4) approximates the low concentration of the standard indicating that the potency lies between 1 and 6 mcg/ml.
Preparation B: Potency of cyanocobalamin has already been assayed and found to lie between 1 and 6 mcg/ml. 4 different dilutions have been assayed (30,000, 50,000, 75,000 and 120,000 fold, nr. 6, 7, 8 and 9 respectively). Of the data obtained the 50,000 fold dilution (nr. 7) yields turbidity values approximating those of standardsets.
Preparations
   C and D: Potency of cyanocobalamin can be estimated. These preparations are assayed in triplicate in the same dilutions.
In case of preparation C: 90,000 fold (nr. 10, 11 and 12).
           preparation D: 250,000 fold (nr. 13, 14 and 15).
Page 16 (page 4 on print-out).
The results of the preparations collected in an easily readable form to be used as data sheet for the Quality Control Department.

Application

As the Dilution and dosage unit starting from one solution prepares the final mixtures in two different concentrations this unit is especially suitable for the performance of two point parallel line assays (Finney, 1952). As soon as conditions have been standardized this highly efficient statistical design can be applied to such an extent that consistent log-dose response curves can be produced. Preferably two points lying on the steepest part of the curve are used for the calculation of the sample content.

If the content of the sample to be assayed can be estimated, the dilution series can be limited to one only, prepared by the Dilution and dosage system. If the content cannot be estimated more dilutions will have to be prepared and assayed (Fig. 10).

The automated system has been applied extensively in routine use for the microbiological turbidimetric, two-point, parallel line assays of vitamins (Berg et al., 1974) and antibiotics. Computer programmes are available for the calculation of contents, 95 % confidence limits and other statistical parameters. The computer print out of a cyanocobalamin assay is shown in Fig. 10. The digilog converter of the automatic turbidity reading unit can be connected electrically with a paper tape-puncher which supplies the turbidity data on paper tape. In Tables 1 and 2 the general outline of the assay methods as applied for the vitamins, thiamin, riboflavin, pyridoxin, cyanocobalamin, panthenol, nicotinic acid, calciumpantothenate, and folic acid, and the antibiotics, tetracycline, chloramphenicol, penicillin, neomycin, streptomycin and dehydrostreptomycin, are given.

Discussion

An average vitamin or antibiotic assay comprises about 140 tubes. In the case of a two-point parallel line assay, these represent 9 reference standards and 26 unknown samples, each in duplicate, at high and low dose level. When performed manually two such assays occupy two technicians with monotonous and tedious work for a whole working day. In addition much special glassware, such as pipettes and volumetric flasks, is needed. Two assays can easily be prepared on one Dilution and dosage unit in one working day, which will need only part of the attention of one technician. The reading of the turbidity can be accomplished in 30 min. So up to four Dilution and dosage units can be used together with one Turbidity reading unit. One experienced technician can easily handle two Dilution and dosage units. The flexibility of the apparatus is shown by the fact that without any special precautions vitamins and antibiotics can be assayed alternately

TABLE 1(a). Vitamin assay test procedure

	Thiamine	Riboflavine	Pyridoxine	Cyanocobalamin
Test organism*	*Lactobacillus viridescens* ATCC 12706	*Lactobacillus casei* ATCC 7469	*Saccharomyces carlsbergensis*** ATCC 9080	*Lactobacillus leichmannii* ATCC 7830
Test broth	Thiamine L.V. medium (Difco)	Riboflavine assay medium (Difco)	Pyridoxine Y medium (Difco)	B12 assay medium (73·3 g/750 ml) (Dano)
Dissolution and extraction	0·001 N HCl	5 ml 25% acetic acid per 10 mg riboflavine	0·06 N HCl	0·0006% KCN solution
Further dilution in	0·001 N HCl	Distilled water	0·06 N HCl	0·0006% KCN solution
High dose level in final sample (μg/ml)	$7 \cdot 5 \times 10^{-3}$	$2 \cdot 5 \times 10^{-2}$	1×10^{-3}	2×10^{-5}
Preparation of inoculum	16 h culture in APT broth (Difco) three times rinsed in and diluted 1:10 with physiological saline	16 h culture in AOAC Lactobacilli broth (Difco) three times rinsed in physiological saline	16 h culture in AOAC Lactobacilli broth (Difco) three times rinsed in and diluted with physiological saline to c. 10^6 cells/ml	16 h culture in AOAC Lactobacilli broth (Difco) three times rinsed in physiological saline
Inoculation of test culture	Culture tubes autoclaved 10 min. 110° C; tube-wise	Inoculum added to test broth bulk solution	Culture tubes autoclaved 10 min. 110° C; tube-wise	Inoculum added to test broth bulk solution
Incubation	16 h at 30° in water bath	18 h at 37° in water bath	16 h at 30° on a shaker	16 h at 37° in water bath.

* Stock stored in liquid N_2 in physiological saline.
** Stock stored in liquid N_2 in Lactobacilli broth AOAC (Difco).

TABLE 1(b). Vitamin assay test procedure

	Calciumpantothenate/ Nicotinic acid	Pantothenol	Folic acid
Test organism*	*Lactobacillus arabinosus* ATCC 8014	*Acetobacter suboxydans*** ATCC 6214	*Lactobacillus casei* ATCC 7469
Test broth	Pantothenate medium USP-Niacin assay medium (Difco)	Panthenol assay medium, Panthenol supplement (Difco)	Folic acid assay medium (Dano)
Dissolution and extraction	Distilled water	In distilled water and hydrolysis in a surplus of 0·1 N NaOH for 30 min at 120°	Distilled water
Further dilution in	Distilled water	Distilled water	Distilled water
High dose level in final sample μg/ml. Preparation of inoculum	$2\cdot5 \times 10^{-2}$ 16 h culture in AOAC Lactobacilli broth (Difco) three times rinsed in physiological saline	$1\cdot25 \times 10^{-1}$ 16 h culture in AOAC Lactobacilli broth (Difco) three times rinsed in and diluted 1:3 with physiological saline	4×10^{-4} 16 h culture in AOAC Lactobacilli broth (Difco) three times rinsed in physiological saline
Inoculation of test culture	Inoculum added to test broth bulk solution	Culture tubes autoclaved 10 min 110°; tube-wise	Incoculum added to test broth bulk solution
Incubation	16 h at 37° in water bath	16 h at 30° on a shaker	16 h at 37° in water bath

* Stock stored in liquid N_2 in physiological saline.
** Stock stored in liquid N_2 in Lactobacilli broth AOAC (Difco).

TABLE 2. Antibiotic assay test procedure

	Tetracyclin	Chloramphenicol	Penicillin	Neomycin	Streptomycin	Dihydro-streptomycin
Test organism*	*Escherichia coli* ATCC 10536	*Escherichia coli* ATCC 10536	*Staphylococcus aureus* 150	*Staphylococcus aureus* 150	*Staphylococcus aureus* 150	*Staphylococcus aureus* 150
Test broth	Tetracylin test broth**	Chloramphenicol test broth***	TGE test broth (double strength) (Difco)	TGE test broth (double strength) (Difco)	TGE test broth (double strength) (Difco)	TGE test broth (double strength) (Difco)
Dissolution and extraction	100 ml 0·1 N HCl	30 ml methanol	30 ml methanol	0·1 Mol phosphate buffer pH 7·9	0·1 Mol Tris buffer pH 8·0	0·1 Mol Tris buffer pH 8·0
Further dilution in	0·067 Mol phosphate buffer pH 7·1	0·067 Mol phosphate buffer pH 6·0	0·067 Mol phosphate buffer pH 6·0	0·1 Mol phosphate buffer pH 7·9	0·1 Mol Tris buffer pH 8·0	0·1 Mol Tris buffer pH 8·0
High dose level in final sample (μg/ml)	5×10^{-2}	1×10^{-1}	$2 \cdot 5 \times 10^{-2}$ IE/ml	1×10^{-1}	1×10^{-1}	6×10^{-2}
Preparation of inoculum	16 h aerated culture in nutrient broth (Difco) diluted with physiological saline 1:2 and 1:3 for *Escherichia coli* and *Staphylococcus aureus* respectively.					
Inoculation of testculture	Tube-wise					
Incubation	3·5 h at 37° in a water bath and on a shaker for *Escherichia coli* and *Staphylococcus aureus* respectively.					

* Stock stored in liquid N_2 in physiological saline.

** medium composition: solution A (9 parts): see **.

Solution B (1 part):
Mannitol (NBC)　　　110 g
Distilled water to　　1000 m

*** medium composition; solution A (9 parts):
Lab. Lemco meat extract (Oxoid)　　1·5 g
Yeast extract (Difco)　　6·5 g
Peptone (Difco)　　5·0 g
Tryptone (Difco)　　10·0 g
Sodiumchloride (USP)　　3·5 g
Dipotassium phosphate (Noury Baker)　　3·7 g
Monopotassium phosphate (Noury Baker)　　1·3 g
Distilled water to　　1000 ml
Solution B (1 part):
Glucose (USP)　　110 g
Distilled water to　　1000 ml

on one apparatus. An ordinary rinsing cycle is sufficient to eliminate any traces of antibiotic that might disturb the next assay.

In most vitamin assays (see Table 1) test broths inoculated with the test organism are used and no special care needs to be applied to keep strict aseptic conditions. In our experience contaminants have no chance to overgrow the inoculum of test organisms. In the case of the antibiotic assays it proved to be of extreme importance to have the inoculum for each tube in an identical physiological state. Hence best results were obtained when the tubes were inoculated tubewise after the preparation of the test cultures.

Determinations of vitamins and antibiotics resulted in mean values with low coefficients of variation, demonstrating that a high degree of precision can be obtained with the described system. The mean vitamin and antibiotic contents of the various vitamin and antibiotic reference standard solutions determined in assays performed at different days varied between 99 and 102%. The coefficients of variance ranged from 3·9 to 9·2%. In the course of five years (Berg et al., 1968) this system has been improved and has now reached its final stage. During this period the system was shown to be a fully reliable system for routine assays.

Acknowledgements

We acknowledge the skilled assistance of H. Meeuwissen during the development phase. We thank J. van Oorschot and G. Witteveen for their important contribution to the construction of the system.

References

BERG, TH. M., PLEUMEEKERS, AGNES, SIMONS, L. M., VIES, J. VAN DER & BEHAGEL, H. A. (1968). The automation of microbiological vitamin assays by means of a modified autoanalyser system. *Ant. van Leeuwenhoek*, **34**, 239.

BERG, TH. M. & BEHAGEL, H. A. (1972). Semiautomated method for microbiological vitamin assays. *Appl. Microbiol.*, **23**, 531.

BERG, TH. M. (1974). Towards automation of microbiological vitamin and antibiotic assays. In *Proceedings 1st Symposium Rapid Methods and Automation in Microbiology*. Wiley & Sons, New York (in press).

DEWART, R. F., NAUDTS, F. & LHOEST, W. (1965). Automation in the microbiological assays of antibiotics. *Ann. N.Y. Acad. Sci.*, **130**, 686.

FINNEY, D. J. (1952). *Statistical method in biological assay*. Griffin, London.

GERKE, J. R., HANEY, T. A., PAGANO, J. F. & FERRARI, A. (1960). Automation of the microbiological assays of antibiotics with autoanalyser instrumental system. *Ann. N.Y. Acad. Sci.*, **87**, 782.

GUALANDI, G. & MORISI, G. (1967). Automation in analysis. I. Apparatus for photometric determination. *Ann. Inst. Super. Sanita*, **3**, 589.

HANEY, T. A., GERKE, J. R., MADIGAN, M. E. & PAGANO, J. F. (1962). Automated microbiological analyses for tetracycline and polyenes. *Ann. N.Y. Acad. Sci.*, **93**, 627.

KUZEL, N. R. & KAVANAGH, F. W. (1971). Automated system for analytical microbiology. I. Theory and design consideration. *J. Pharm. Sci.*, **60**, 764.

KUZEL, N. R. & KAVANAGH, F. W. (1971). Automated system for analytical microbiology. II. Construction of system and evaluation of antibiotics and vitamins. *J. Pharm. Sci.*, **60**, 767.

MCMAHAN, J. R. (1965). A new automated system for microbiological assays. *Ann. N.Y. Acad. Sci.*, **130**, 680.

PLATT, T. B., GENTILE, J. & GEORGE, M. J. (1965). An automated turbidimetric system for antibiotic assay. *Ann. N.Y. Acad. Sci.*, **130**, 644.

SHAW, W. H. C. & DUNCOMBE, R. E. (1963). The continuous automatic microbiological assay of antibiotics. *Analyst*, **88**, 694.

TSUJI, K., GRIFFITH, D. A. & SPERRY, C. C. (1967). Automated turbidimetric bioassay readout instrument using a multiple flow-cell system. *Appl. Microbiol.*, **15**, 145.

An Automatic Dilution Apparatus for use in Vitamin and Antibiotic Assay

G. H. Palmer and P. Hamilton

Beecham Research Laboratories, Walton Oaks, Dorking Road, Tadworth, Surrey, England

Microbiological assay either by tube or agar plate methods can be divided into three activities:

1. Dilution of sample to give concentrations of active constituent which fall within the range of standards for a given assay organism.

2. Application of the diluted solution to the assay system, i.e. tubes or cups for the agar plate method.

3. Measuring growth or inhibition of growth (or parameters of growth) of the assay organism by the material under test and calculating the potency of the sample.

The dilution of a sample may require considerable quantities of glassware, it is time consuming, tedious and, in laboratories with a large throughput of samples, can involve a considerable working area. In addition all glassware has to be adequately cleansed which may be a critical factor for satisfactory vitamin assay.

A system of automated dilution coupled with disposable-collection vessels for the diluted solutions would therefore be of considerable advantage in a laboratory carrying out microbiological assays.

A suitable automatic diluting system should have the following characteristics:

1. It should be compact and require no specialized working area.

2. It should be simple in operation and require minimal expertise by the operator.

3. It should be accurate and give reproducible dilutions.

4. It should be capable of use with disposable receiving vessels thus avoiding the need for any cleaning processes.

In addition:

(a) The volume of sample required for dilution should be kept to a minimum since there may be a restriction in quantity of sample available for analysis (cf pharmaceutical ampoules).

(b) The volume of diluted sample should be sufficient for the requirements of the assay system. In practice vitamin tube assays require the larger volumes of solutions and 40 ml satisfies this requirement in our laboratory.

(c) A comprehensive range of dilutions should be available though the use of combinations of dilution steps reduces considerably the number of dilution ratios necessary.

(d) Carryover between dilutions should be minimal.

The following are the details of a dilution system which has been devised to satisfy these criteria and which is in current use in our laboratory for the microbiological assay of antibiotics, amino acids and vitamins.

Description of system

The dilution system described is a modified version of a commercially available, variable diluter (Hook and Tucker Ltd) coupled with the use of disposable plastic collection vessels. (Luckhams Ltd, Burgess Hill, Sussex). The diluter was produced to our specifications by the manufacturer and differs fron the standard diluter as follows:

1. Dilutions are set on a series of stops which are preferred to a continuous dilution range because it avoids any subjective assessment of the dilution setting by the operator.

2. The total volume dispensed is 40 ml in order to satisfy solution requirements for assays.

Operation

To operate the diluter (Fig. 1), the dilution required is selected on the dial by rotating the dilution selector. The sample to be diluted is held under the sampling tube and the switch depressed (either hand or foot switch). The sample syringe then aspirates the required sample volume into the sampling tube (note that the sample volume never enters the sample syringe). On depressing the switch a second time, the sample volume is delivered together with the diluent through the sample tube, the diluent also serving to flush any residual sample from the walls of the sampling tube.

All solutions are collected into disposable plastic cups, mixing taking place as both the sample volume and diluent are delivered into the cup, i.e. external to the machine.

The lamps 1 and 2 serve to indicate when a particular part (uptake or delivery) of the dilution cycle has been completed.

Dilutions

The dilution ratios available on this machine are $\frac{1}{4}$, $\frac{1}{5}$, $\frac{1}{10}$, $\frac{1}{17}$, $\frac{1}{20}$, $\frac{1}{40}$, and $\frac{1}{100}$ which from our experience provide the majority of dilutions required, but if particular dilutions other than those specified are required then these settings can easily be altered.

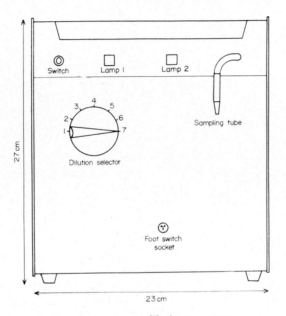

FIG. 1. Automatic dilution apparatus.

Laboratory Evaluation

Precision, accuracy and reproducibility

Table 1 shows the mean dilution ratio, fiducial limits (P = 0·05) and % fiducial limits for each of four operators carrying out five replicate operations for each dilution ratio. The final column shows the mean dilution for the four operators.

To obtain this data N HCl was diluted in distilled water and 10 ml of the diluted solution was titrated with 0·01 N NaOH to pH 7·0 using an automatic titrimeter (model 24 Electronic Instruments Ltd).

Carryover

In a study using the $\frac{1}{4}$ dilution ratio and concentrated vitamin solutions,

TABLE 1. Operators—Dilutions 1/x

Nominal dilution	A	B	C	D	Overall
1/4	4·00	4·08	4·14	4·05	4·07
Limits	3·99–4·02	4·05–4·10	4·10–4·17	4·03–4·06	
% Limits	± 0·29	± 0·65	± 0·69	± 0·49	
1/5	5·08	4·99	5·18	4·88	5·00
Limits	5·04–5·14	4·94–5·04	5·12–5·23	4·83–4·92	
% Limits	± 1·00	± 1·10	± 1·03	± 0·85	
1/10	10·01	10·06	10·55	10·34	10·24
Limits	9·96–10·06	9·89–10·22	10·50–10·60	10·27–10·42	
% Limits	± 0·55	± 1·60	± 0·53	± 0·72	
1/17	16·50	16·75	16·90	17·32	16·87
Limits	16·41–16·58	16·60–16·90	16·81–17·06	17·04–17·61	
% Limits	± 0·60	± 0·89	± 0·58	± 1·66	
1/20	19·52	19·52	20·85	20·35	20·06
Limits	19·33–19·71	18·80–20·28	20·46–21·25	20·22–20·50	
% Limits	± 1·00	± 4·06	± 1·87	± 0·68	
1/40	39·85	42·00	41·30	40·01	40·79
Limits	39·82–39·90	—	40·9–41·6	38·33–41·98	
% Limits	± 0·10	—	± 0·79	± 4·56	
1/100	99·80	102·90	99·2	102·20	101·0
Limits	—	99·7–106·40	91·1–101	97·9–106·9	
% Limits	—	± 3·27	± 2·22	± 4·39	

less than 1% of the sample volume remained in the sample tube after completion of the dilution cycle.

Factors necessary to ensure satisfactory machine operation

1. Ensure that all the air is removed from the machine at the start of each session by operating the machine with diluent only over several dilution cycles; never use refrigerated buffer as this will release dissolved gases when it warms up in the machine.
2. Use sterile diluent to minimize the risk of microbial contamination within the autodiluter. This procedure also reduces contamination on assay plates.
3. After taking up sample, wipe (with an absorbant tissue or filter paper) the tip of the sampling tube to remove any sample on the outside of the tube; any adhering liquid is taken into the sampling tube at the commencement of the delivery cycle and would give inaccurate dilutions.
4. Hold the receiving vessels at such an angle and height so as to prevent splashing during the delivery cycle.
5. Always operate the machine for one cycle with diluent only if an occasion arises where a dilution of a very dilute solution is required immediately after dilution of a relatively concentrated solution or between different samples. It is normal practice to carry out dilutions sequentially for a given sample. Any effects due to carryover are thus minimized.

Discussion

The dilution system described has been in use in our laboratory for 2 years; it is inexpensive, highly satisfactory in reducing effort in sample dilution and, although data for a direct comparison with conventional diluting procedures are not available, we feel that this system offers distinct advantages over the latter, specifically with regard to precision, reproducibility and lack of tedium. In conclusion it is emphasized that the use of disposable plastic cups as collection vessels is an essential component of this system and only by their use can the maximum potential of the diluter be achieved.

Application of Automated Assay of Asparaginase and other Ammonia-releasing Enzymes to the Identification of Bacteria

SHOSHANA BASCOMB AND C. A. GRANTHAM*

Department of Biochemistry, Imperial College of Science and Technology, London SW7 2AZ, England

The impetus to study L-asparaginase stems from the demonstration of its therapeutic effect in treatment of experimental and human cancer. The subject has been studied very extensively and reviewed recently (Capizzi, Bertino and Handschumacher 1970). Microorganisms that lend themselves to large scale production have proven the best source for L-asparaginase; enzymes from *Escherichia coli*, *Erwinia carotovora* and *Serratia marcescens* have been purified and used successfully in clinical trials (Cooney and Handschumacher, 1970). L-asparaginase was found effective against many experimental tumors, but its effectiveness spectrum in human therapy is limited. Research with L-asparaginase has shown the first exploitable metabolic difference between certain neoplastic cells and normal ones. This started the exploration of enzyme therapy, and stimulated our interest in the potential use of other microbial enzymes for cancer therapy.

The application of automated continuous flow methods to study bacterial metabolism started in the field of microbiological assay of antibiotics (Gerke *et al.*, 1960). The method employs measurement of turbidity or CO_2 for the determination of bacterial growth, further developments in this field have been reviewed by Gerke and Ferrari (1968). Dealy and Umbreit (1965) measured automatically β-galactosidase, acid and alkaline phosphatase activities, as well as protein, DNA, RNA and turbidity of growing cultures of *E. coli*. Leclerc (1967) used CO_2 production for estimation of glutamic acid decarboxylase activities of 230 strains of bacteria, and Trinel and Leclerc (1972) showed the value of the technique in classification and identification. Bettelheim, Kissin and Thomas (1970) described an automatic technique for determination of ammonia produced by bacteria based on the Berthelot reaction, and used it for measuring

* Present address: Research and Development Division, G. D. Searle & Co. Ltd., Lane End Road, High Wycombe, Bucks, HP12 4HL, England.

L-asparaginase and L-glutaminase activities of 1000 bacterial strains. Wade, Robinson and Phillips (1971) used automated determination of ammonia based on the Nessler reaction to measure L-asparaginase and L-glutaminase activities of 200 bacterial strains, and found that both enzymes were widely distributed among bacteria.

We tried to adapt the above techniques for measurement of ammonia released from a variety of substrates and found that the Berthelot reaction was not specific for ammonia. We had to develop a modification of the Nessler reaction for use with our samples. This contribution describes an automated method for measurement of different ammonia-releasing enzymes, its application to cell suspensions, and its potential use in bacterial identification.

Materials and Methods

Reagents

Solutions were made from BDH analytical reagents unless otherwise stated. *Alkaline copper* solution was prepared daily by mixing 5 ml sodium potassium tartrate (2% w/v)—kept at 5°—with 5 ml $CuSO_4.5H_2O$ (1% w/v) and 200 ml of carbonate buffer (10% (w/v)Na_2CO_3 and 1% (w/v) NaOH) in the order described.
Folin-Ciocalteu (BDH) reagent was diluted daily 1:9 with distilled water.
Hydrochloric acid—0·01N and 1N solutions were prepared.
Nessler reagent was prepared by a method described by Koch and Mc-Meekin (1924). A concentrate was prepared by dissolving 30 g KI in 20 ml distilled water followed by 22·5 g iodine. Thirty g of mercury were added to the above solution, shaken until a straw yellow colour appeared, decanted and the supernatant tested for free idoine with 1% (w/v) soluble starch solution. If positive, it was diluted with water to 200 ml, and 975 ml of NaOH (10% w/v) added. This reagent was allowed to stand in the dark for 24 h to clear. A 10% dilution in water of the concentrated reagent was prepared afresh daily.
Potassium borate buffer—a 0·05 M (pH 8·0) solution was used.
Standards containing both ammonium chloride and Bovine Albumin Fraction V (Sigma Chemicals Co.) were prepared in 0·05M borate buffer (pH 8·0) in the concentrations shown in Table 1. A few drops of chloroform were added to each standard solution to prevent growth of contaminants.
Substrates (10 mM, except adenine and adenosine 5mM) were prepared in 0·05 M borate buffer, the pH adjusted to 8·0, and kept at 5° for 4–6 weeks. A fresh solution was prepared when free ammonia was detected in the

control channel. Solutions of L-asparagine and L-glutamine were prepared daily as they deteriorated quickly. Purine and pyrimidine bases and their sugar derivatives were obtained from Koch Light Laboratories. The list of substrates is given in Table 1.

Media

Corn Steep Liquor (CSL) (Garton & Son Ltd, York Place, Battersea, London SW11) was prepared according to the method of Roberts, Burson & Hill (1968).
Glycerol Asparagine (GA) medium contained (/l): glycerol, 10 g; L-asparagine, 1 g; K_2HPO_4, 1 g; $FeSO_4.7H_2O$, 1 mg; $MnCl_2.4H_2O$, 1 mg, and $ZnSO_4$, 1 mg pH 7·0–7·4.
Nutrient Broth 'CMI' (NB) was obtained from Oxoid Ltd., London.
Yeatex Lactic Acid 'YLA' medium contained (/l): Yeatex granules (Bovril Food Ingredients, Bovril Ltd., Wellington Road, Burton on Trent, Staffordshire), 20·6 g; lactic acid, 11·25 g, and K_2HPO_4, 2·5 g; pH 7·0.
Yeatex Glutamate (YG) was prepared according to Buck *et al.* (1971) and contained (/l): Yeatex granules, 40 g, and sodium glutamate, 17 g; pH 7·0.

Organisms

Sixty strains from the departmental culture collection were examined. The identity of each strain was confirmed on the basis of conventional testing following the methods and characterisations described by Cowan and Steel (1965) and Bascomb *et al.* (1973).

Preparation of bacterial suspension for automated assay

Bacteria are inoculated into 9 ml peptone water (Oxoid) and incubated at 37° for 8 h. The entire contents of the tube is then transferred aseptically to 100 ml of CSL in a 500 ml conical flask. The flasks are incubated for 18 h at 37° on a rotary shaker with a radius of gyration of 45 mm and a speed of 200 gyration/min. Cells are harvested by centrifugation of 80 ml for 10 min at 17000 rpm in an MSE High Speed 18 centrifuge (Measuring & Scientific Equipment Ltd., 25–28 Buckingham Gate, London SW1) at 4°, washed once in 0·05 M potassium borate buffer (pH 8·0) and resuspended in 20 ml of the same buffer. These concentrated cell suspensions are assayed immediately or kept at −20°.

Ultrasonication

Concentrated cell suspensions are disrupted by ultrasonication for 8×15 sec in a 150 W MSE Ultrasonic Disintegrator (amplitude setting 8 μm) the sample chamber cooled to 5°; 60% kill is generally achieved. Disrupted

cells are diluted to 0·15–0·3 of the original broth concentration and
supplied in 50 ml aliquots to probe 2 of the Auto-Analyzer (Fig. 1).

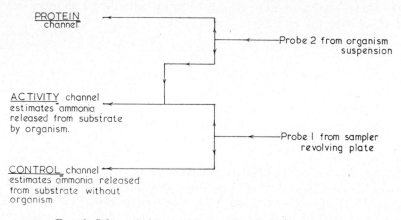

FIG. 1. Schematic layout of the 3 channel analytical system.

Automated assay

The automated assay system is based on the continuous flow principle
and consists of three analytical channels running simultaneously (Fig. 1).
One channel measures ammonia present in the substrate (control channel);
the second, ammonia released from the substrate by the bacterial suspension (activity channel), and the third measures the protein content of the
bacterial suspension. The chemical basis and manifold layout of the two
reactions are described first, followed by the complete detailed manifold.

Ammonia estimation

This is based on colorimetric determination of the yellow colour produced
by the Nessler reaction. The schematised manifold is shown in Fig. 2.
The substrate stream from probe 1, connected to the revolving sampler
plate, is split into two streams, both are segmented by air, and joined in
channel 1 to a stream of buffer, and in channel 2 to a sample of bacterial
suspension. Each stream is mixed in a single mixing coil (SMC) and
pumped into a 13 min delay coil in an oil bath kept at 37°. The reaction
mixture is then passed through one set of dialyzer plates (2 min delay).
Ammonia present in the sample diffuses through the dialyzer membrane
(Technicon, Standard) into an air-segmented recipient stream of 0·01 N
HCl which is then treated with 10% Nessler reagent, mixed in a SMC,
passed through a 15 mm flow cell in a colorimeter and the optical density

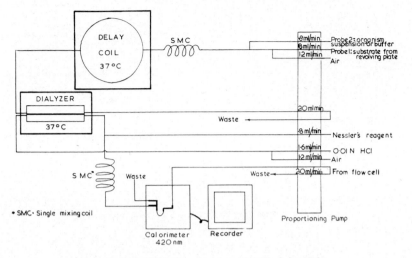

FIG. 2. Manifold of ammonia estimation channels.

(420 nm) registered on one channel of a two pen recorder. The introduction of a Nessler reagent to a dialyzate from a concentrated bacterial sample caused precipitation which was cleared between runs with a 1 N HCl solution introduced via the Nessler line. Standards and substrate samples are supplied to the analytical system in the sequence (Table 1) arranged on the sampler revolving plate. Total assay time is 17 min. The traces of the two ammonia channels were synchronised and both appeared on the same chart (Fig. 5). Sensitivity range is $0.3–2.4$ μM NH_3/ml. Assuming reaction time of 15 min $(13 + 2)$, these ammonia values correspond to $0.02–0.16$ International Units (i.u.) of enzyme activity using the term as defined for L-asparaginase, namely μM NH_3 released/ml/min.

Protein estimation

This is based on the Lowry *et al.* (1951) method (Fig. 3). The air-segmented sample stream is mixed with alkaline copper reagent in a double mixing coil (DMC) and treated in a second DMC with 10% Folin-Ciocalteu reagent (BDH). The mixture is de-bubbled, passed through a 15 mm flow cell in a colorimeter and the optical density (620 nm) registered on a Vitatron recorder set to logarithmic scale. Samples were supplied to the analytical system from the bacterial suspension via probe 2 and provided a protein assay for the bacterial suspensions incubated with each substrate. Total assay time is 10 min. At the beginning of each working day the response of standards was achieved by transfer of probe 2 to the revolving

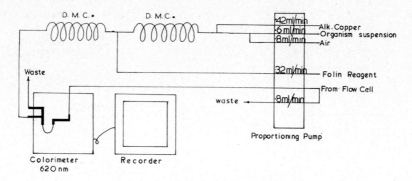

FIG. 3. Manifold of protein estimation channel.

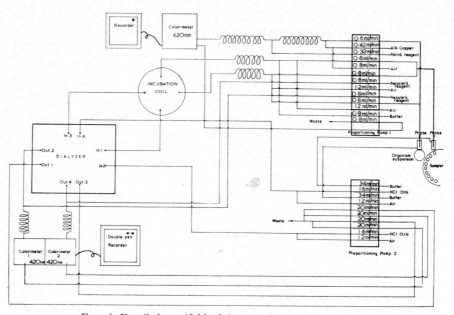

FIG. 4. Detailed manifold of the complete analytical system.

sampler plate and assay of the first 7 samples. Probe 2 was then returned to its normal position and the sampler plate turned back to sample 1 (Fig. 6). Sensitivity range is 0·015–0·500 mg protein/ml with Bovine Albumin Fraction V as standard.

The whole analytical system (Fig. 4) is run at the rate of 50 samples/h with borate buffer as the wash (sample: wash ratio, 1:1).

Equipment

The following Technicon (Technicon Instruments Co. Ltd., Hamilton Close, Houndmills, Basingstoke, Hants) Auto-Analyzer I modules are used: a Sampler II modified to carry a double probe (Pollard and Waldron, 1967) so that organism suspensions are aspirated concomitantly with the substrates, sample and organism streams are both intercalated by wash; two Proportioning Pumps II; an Oil Bath containing 2×20 ft glass coils I.D. 2·4 mm; a Dialyzer with two sets of dialysis plates; two Colorimeters each with a 15 mm flow cell; a Double Pen Recorder. A Bausch and Lomb (Applied Research Laboratories Ltd., Windsgate, Luton, Beds.) Spectronic 20 Colorimeter converted to continuous flow (15 mm flow cell) was connected to a Vitatron recorder (Fison Scientific Apparatus Ltd., Bishop Meadow Road, Loughborough, Leicestershire).

Computation

Peak heights were read by observation, the values punched on cards and calculated in a CDC 6600 computer. The computer printed for each sample the following values:

A) Enzyme activity i.u.*/ml $=$

$$\frac{\mu\text{M } NH_3/\text{ml of activity peak} - \mu\text{M } NH_3/\text{ml of control peak}}{\text{Reaction time (15 min)}} \times \text{Dilution factor}$$

B) Protein content mg/ml.

C) Specific enzyme activity i.u./mg protein $=$ A/B.

Results

Preliminary experiments were done on suspensions of bacteria from different taxonomic groups using as substrates L-amino acids as well as adenine, cytosine, guanine and their derivatives (30 in all). Some substrates interfered with the Nessler reaction, some were unstable under the conditions of the assay, and some gave no positive results with any of the tested strains. Fifteen substrates were therefore selected for routine assays. It was noticed also that, owing to insufficient sample separation, false positive results occurred with substrates immediately following samples with high enzymatic activity. Buffer samples introduced between substrates, acting as additional washes, minimized this carry over; L-asparagine which yielded very large responses with many strains was followed by two buffer samples. Peak identification was further helped by inclusion of an

* Using the term as defined for L-asparaginase.

TABLE 1. Activities of ammonia-releasing enzymes calculated for a typical SEP of *Proteus morganii A114**

Arrangement on revolving sampler tray		Enzyme activity 100 × i.u./ml (A)	Protein mg/ml (B)	Specific activity 100 × i.u./mg protein (C)
Cup No.	Content**			
	μM NH₃/ml mg Alb/ml			
1	0·03 + 0·015 ⎤			
2	0·60 + 0·030 ⎥			
3	1·20 + 0·060 ⎥ (Standards)			
4	1·80 + 0·120 ⎥			
5	2·40 + 0·240 ⎥			
6	Buffer + 0·500 ⎦			
7	Buffer 2	6	2·61	2
8	adenine	4	2·62	1
9	Buffer 3	7	2·62	3
10	adenosine	10	2·60	4
11	Buffer 4	7	2·60	3
12	L-alanine	6	2·59	3
13	Buffer 5	6	2·61	2
14	L-asparaginase	40	2·61	15
15	Buffer 6	20	2·64	8
16	Buffer 7	7	2·57	3
17	L-aspartic acid	7	2·60	3
18	Buffer 8	5	2·55	2
19	cytidine	7	2·60	3
20	Buffer 9	7	2·55	3
21	Marker	–	2·62	–
22	Buffer 10	6	2·56	2
23	L-glutamic acid	5	2·56	2
24	Buffer 11	5	2·57	2
25	L-glutamine	14	2·55	5
26	Buffer 12	10	2·61	4
27	L-lysine	4	2·56	1
28	Buffer 13	5	2·55	2
29	L-methionine	23	2·60	9
30	Buffer 14	8	2·60	3
31	L-β-phenylalanine	36	2·61	14
32	Buffer 15	9	2·57	4
33	L-threonine	31	2·60	12
34	Buffer 16	10	2·61	4
35	L-tryptophane	11	2·63	4
36	Buffer 17	5	2·62	2
37	L-serine	15	2·55	6
38	Buffer 18	10	2·64	4
39	urea	113	2·63	43
40	Buffer 19	15	2·63	5
41	Marker	–	2·63	–

* Based on Figs 5 and 6. **Substrates, 0·01 M, except adenine and adenosine, 0·005M; all in 0·05 M buffer. Buffers, all 0·05 M-borate (pH 8·0).

ammonium standard in the middle and at the end of each tray set as marker samples. In routine assay each organism was tested with 41 samples comprising 15 substrates, 7 standards and 19 buffer samples, in the sequence and concentrations shown in Table 1. The time taken for testing one organism was 48 min. All results were available 65 min after aspiration of the first sample; 6–7 strains were tested during a normal working day (> 800 tests).

The enzymatic activities of whole cells of *Proteus morganii A114* are shown in Fig. 5; visual examination shows that ammonia was liberated from the following substrates in decreasing order: urea, L-asparagine, L-β-phenylalanine, L-threonine, L-methionine, L-serine and L-glutamine. Less ammonia was liberated from L-tryptophane, adenosine, cytidine and L-aspartic acid. The recorder trace of a routine run (Fig. 5), showing the enzymatic activities present in one bacterial suspension, was termed "Specific Enzyme Profile" (SEP). Protein content of the organism suspension samples incubated with the different substrates is shown in Fig. 6 and is virtually constant. The activity, protein content and specific activity calculated for each sample are shown in Table 1. The amount of endogenous ammonia released from the cell suspension is denoted by "activity" of the buffer samples. The specific activities of the enzymes noted above are 2–20-fold greater than the values of buffer samples unaffected by carry over, e.g. buffer samples 8 and 11.

Effect of growth medium on Specific Enzyme Profile (SEP)

Strains of *Acinetobacter* sp., *Aeromonas formicans*, *A. hydrophila*, *Bacillus* sp., and *Pr. morganii* were grown overnight in shaken flasks containing CSL, GA, NB, YG or YLA. Cells were harvested and assayed for SEP. Generally speaking, protein content was greatest in cells grown in CSL, slightly less in YG, YLA or NB and much less in GA. SEP's of the same strain grown in different media were similar in the number of enzymes showing activity and in their relative magnitudes. However, quantitative differences were noticed (Fig. 7); greatest activities were obtained from most strains when grown in CSL. A qualitative difference was noticed with strains of *Aeromonas* which produced more L-aspartic acid deaminase and L-asparaginase when grown in CSL but more cytidine deaminase when grown in NB. Cultures for routine assays were grown in CSL.

Effect of cell disruption

Wade *et al.* (1971) suggested that slightly higher L-asparaginase values were obtained from whole organisms than from disrupted preparations;

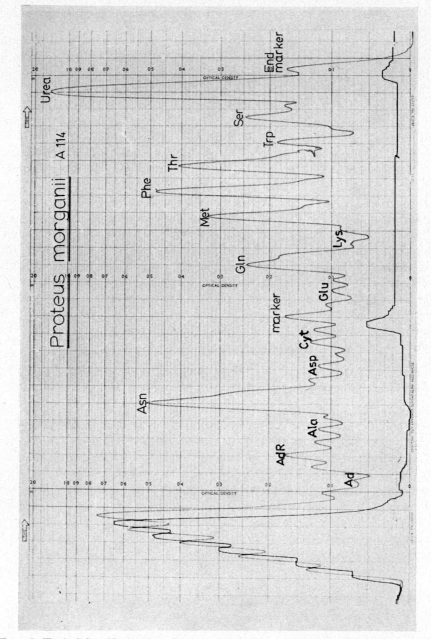

FIG. 5. Typical Specific Enzyme Profile of *Proteus morganii* A114. Chart-traces show ammonia released from samples in the sequence given in Table 1. Upper trace: Activity channel, and lower trace: Control channel.

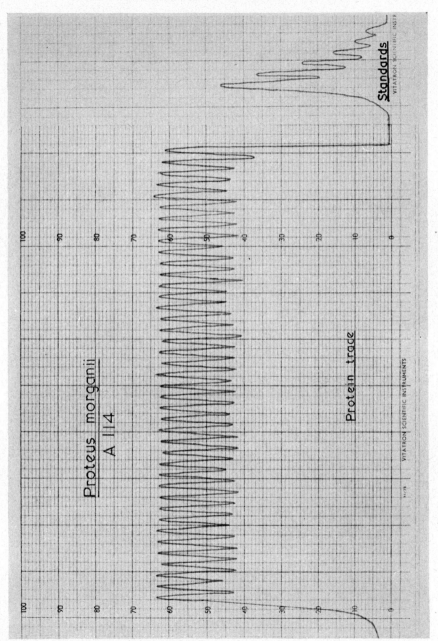

FIG. 6. Protein content of repeat aliquots of the suspension of *Proteus morganii A114* reacting with the 41 cups on revolving sampler plate and giving the enzymatic activities shown in Fig. 5.

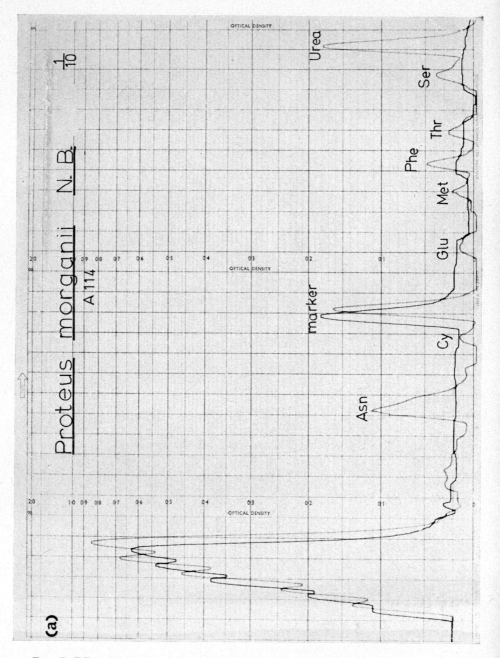

FIG. 7. Effect of growth medium on Specific Enzyme Profile of *Proteus morganii A114* (a), SEP when grown in Nutrient Broth, and (b), SEP when grown in Corn Steep Liquor.

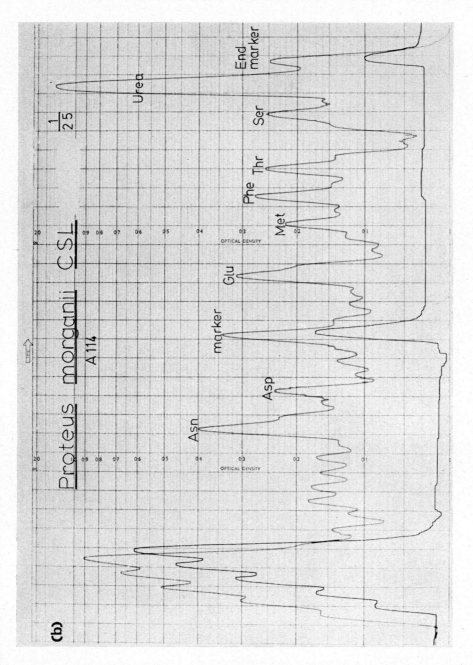

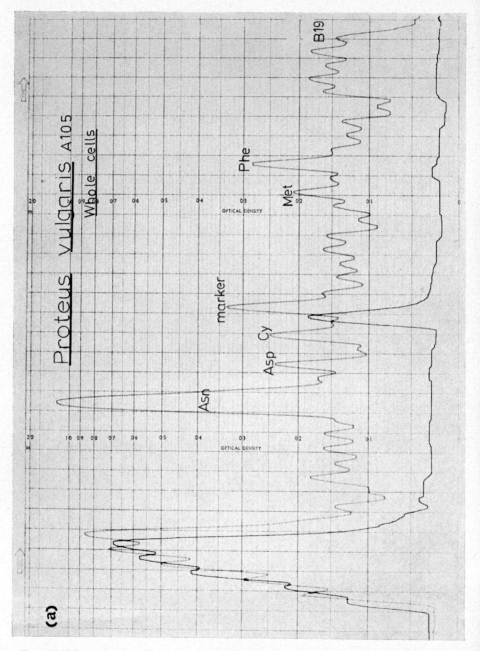

FIG. 8. Effect of ultrasonication on Specific Enzyme Profile of *Proteus vulgaris A105*: (a), whole cells, and (b), after 8 × 15 sec ultrasonication at an amplitude of 8 μm.

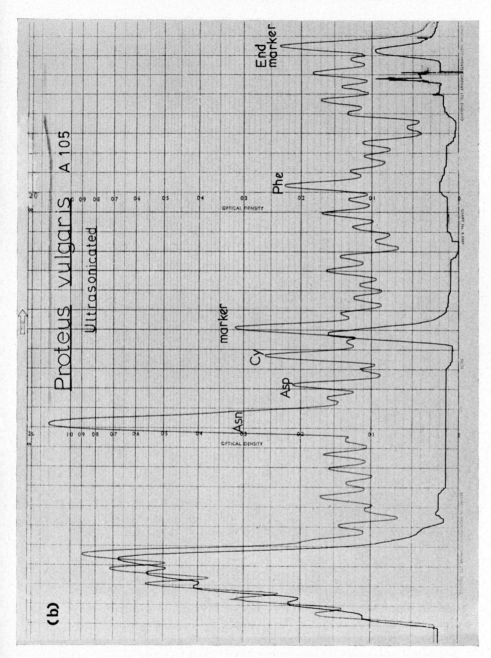

only one strain out of 9 exhibiting greater activity with disrupted cells. SEP's of *Proteus vulgaris A105* before and after ultrasonication (Fig. 8) show that most enzyme activities were only slightly lower in disrupted preparations. Enzymes of other strains, L-tryptophanase of *Pr. morganii* in particular, disappeared completely after ultrasonication; but disrupted *Klebsiella aerogenes* cells had 5 times more cytosine arabinoside deaminase specific activity than whole cells (0·1:0·5 i.u./mg protein respectively), suggesting absence of the necessary permease. Routine assays were done on ultrasonicated samples.

Reproducibility

The specific enzyme activities of 8 strains of *Pseudomonas aeruginosa* are given in Table 2. All strains possess similar levels of L-asparaginase and

TABLE 2. Specific activities of ammonia-releasing enzymes of 8 strains of *Pseudomonas aeruginosa*
Specific activities 100 × i.u./mg protein

	Strain No.							
Substrate	P11	P13	P14	P15	P17	P19	P20	P21
adenine	3	3	2	3	4	5	4	0
adenosine	1	2	0	1	2	0	1	0
L-alanine	6	5	5	4	5	5	6	3
γ-amino butyric acid	0	2	5	5	1	5	1	0
L-asparagine	15	11	13	10	15	9	16	11
L-aspartic acid	3	4	3	3	3	7	5	1
cytidine	0	2	0	0	2	3	0	0
L-glutamic acid	2	4	2	3	3	4	2	0
L-glutamine	12	9	11	8	20	8	8	8
L-lysine	2	3	1	3	4	4	2	0
L-methionine	1	3	1	3	4	4	2	0
L-β-phenylalanine	1	3	1	2	3	5	1	0
L-threonine	2	4	3	4	6	5	3	0
L-tryptophane	1	0	0	0	0	2	0	0
L-serine	6	7	4	5	6	9	7	0
urea	1	4	1	2	3	5	4	1
buffer	2	1	2	3	3	3	3	0

L-glutaminase activities. The values vary between 0·09 and 0·16 i.u./mg protein for L-asparaginase and 0·08-0·20 i.u./mg protein for L-glutaminase. These two amidases are clearly of diagnostic importance in this species. Enzymes releasing ammonia from L-serine, L-alanine and L-threonine appear in fewer strains, have lower activity and are therefore less reliable as diagnostic characters for this species.

Specific enzyme profiles of different species

To test the usefulness of this system for identification purposes, the SEP's of 5 strains of each of the 6 recognized species of the tribe Proteeae (Edwards and Ewing, 1972) were prepared. The SEP's of all strains of 5 species were very similar and quite different from those of other species. The SEP's of *Pr. vulgaris* were less uniform. Representative SEP's for all 6 species are shown in Fig. 9 and summarized in Table 3; L-asparaginase

TABLE 3. Activities of ammonia-releasing enzymes of organisms of the tribe Proteeae

Substrate	Proteus mirabilis	Proteus morganii	Proteus rettgeri	Proteus vulgaris	Providencia alcalifaciens	Providencia stuartii
adenine	—	—	—	—	—	—
adenosine	—	tr	—	—	tr	tr
L.-alanine	—	tr	—	—	—	tr
L-asparagine	+++	++	++	+++	+++	++
L-aspartic acid	—	tr	+	—	tr	+
cytidine	+++	tr	+	++	tr	tr
L-glutamic acid	—	tr	—	—	tr	—
L-glutamine	tr	tr	tr	tr	tr	—
L-lysine	—	—	+	—	—	tr
L-methionine	++	+	+	+	tr	+
L-β-phenylalanine	+++	++	++	++	+	++
L-threonine	—	+	—	—	—	—
L-tryptophane	tr	—	tr	—	—	—
L-serine	+	+	tr	+	tr	—
urea	—	+++	++	—	—	—

Scoring: tr, 0·01–0·04; +, 0·05–0·09; ++, 0·10–0·18, and +++, > 0·18 i.u./mg protein

and L-β-phenylalanine deaminase activities are shown by all strains and could be used as diagnostic features for the tribe. Urease, and enzymes releasing ammonia from L-serine, L-methionine and L-threonine appear only in some species and are of diagnostic value within this tribe.

Discussion

By reason of the use of ultrasonicated suspensions and the short reaction time (15 min), our system measures only activities of enzymes present in the cell. As cells prepared for assay must be grown in a medium, it is obvious that the ingredients of this medium will determine the phenotypic

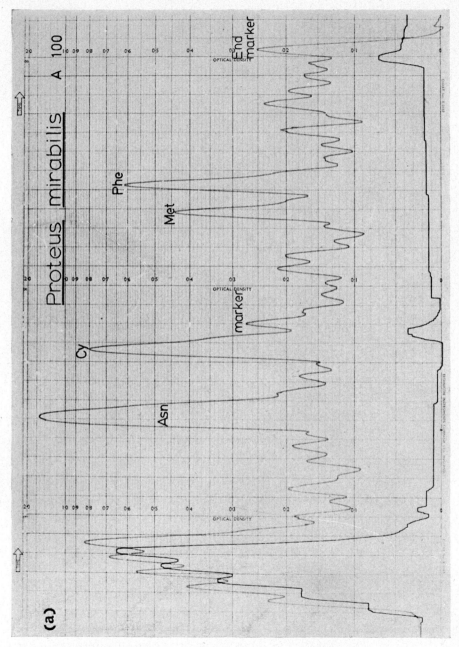

FIG. 9. Specific Enzyme Profile of the tribe Proteeae.

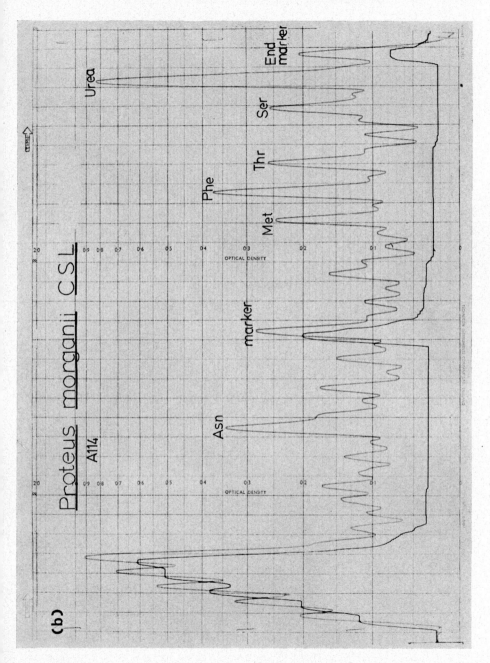

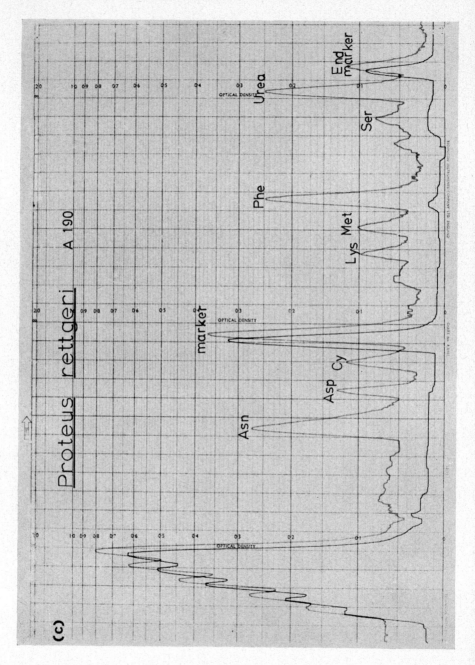

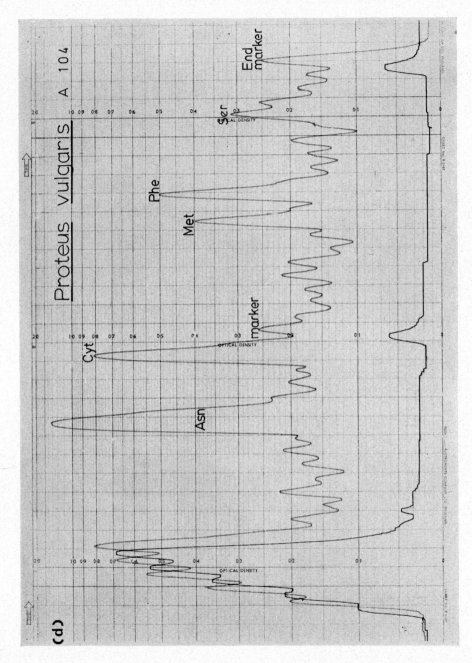

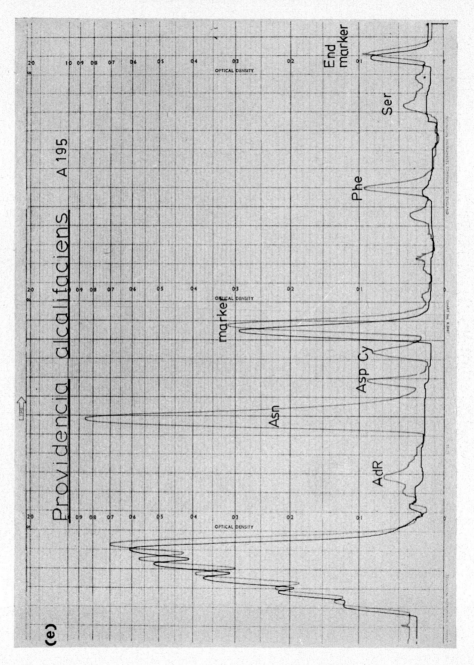

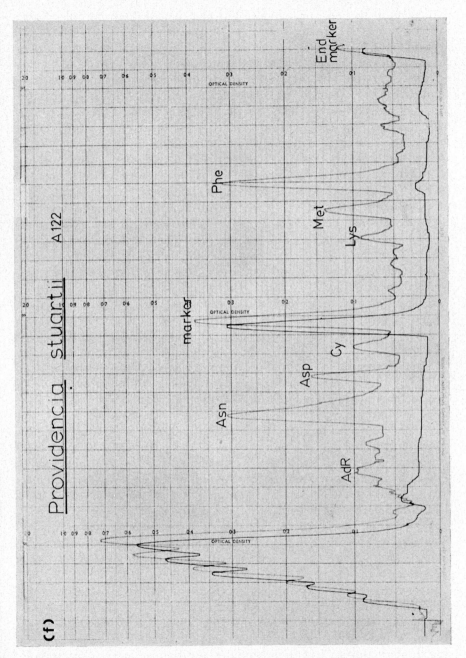

expression of the bacterial genom. The fact that SEP's from cells grown in CSL rich in certain amino acids, NB rich in peptides or GA with L-asparagine as sole source of ammonium were fairly similar, suggests that most of the enzymes detected are constituative. It follows that results obtained with the analytical system may differ from those obtained by traditional tests especially where, in the latter, time allows for enzyme induction, e.g. decarboxylases, "sugars". For example, urease activity, very characteristic of all *Proteus* spp. by traditional testing, was present only in strains of *Pr. morganii* and *Pr. rettgeri* but not in *Pr. mirabilis* or *Pr. vulgaris* (Fig. 9 b, c, a, d); suggesting that in the last two species urease is an induced enzyme. The fact that the Patho-tec paper strip technique (W. R. Warner & Co. Ltd., Eastleigh, Hampshire), recommends 2 h incubation for detection of urease activity but not for the detection of L-β-phenylalanine deaminase, supports the assumption that urease is an induced enzyme. It may well be that other conventional diagnostic features are also of the induced type and that different features will have to be used in automated identification; further testing of more species should indicate which these are.

Experimental evidence showed that ultrasonication left many enzymes unaffected (Fig. 7), but some were destroyed while the activity of others increased 5-fold. Our reasons for doing the routine assays with disrupted cells were:

(A) Ultrasonicated preparations showed better sample separation, and

(B) as our main object is the extraction and purification of enzymes, those sensitive to cell disruption present greater difficulties and are of less immediate interest.

A workable identification system requires that each taxonomic unit will show a unique pattern of results and that such a pattern will be reproducible for all members of the unit. The uniqueness of each profile (SEP) of members of the tribe Proteeae, as displayed by the presence of individual enzymes and by their relative magnitudes, is quite striking (Fig. 9 and Table 3). The reproducibility within one strain on different occasions was very good (Figs 5, 7b, 9b); within species reproducibility was also good (Table 2), with the exception of *Pr. vulgaris* (Figs. 8b and 9d). The SEP system offers an easy and quick method for the identification of these species.

In comparison with conventional methods of testing and identification, the automated system reduces overall testing time by 3–12 days and lessens considerably the labour involved in media preparation, inoculation and reading of test results. Moreover, extension of automation to data processing through a direct link to a computer and the use of an identification programme, similar to that described by Willcox *et al.* (1973), will reduce the time and labour even further and positive identification could be achieved within one hour of obtaining the disrupted cell suspension. Finally, development of additional analytical systems, based on measurement of other metabolic products will help identification of species which show little diversity with regard to deaminating enzymes, and will help to form a complete picture of the metabolic activities of microorganisms.

Acknowledgements

We thank Professor Sir Ernst Chain, F.R.S., for his interest, Dr K. A. Bettelheim for providing the cultures, and Mr K. Blanshard for the computer programme, and advice on electronics. The work was supported by the Cancer Research Campaign.

References

BASCOMB, S., LAPAGE, S. P., CURTIS, M. A. & WILLCOX, W. R. (1973). Identification of bacteria by computer II. Identification of reference strains. J. gen. Microbiol., **77**, 291.

BETTELHEIM, K. A., KISSIN, E. A. & THOMAS, A. J. (1970). An automated technique for the determination of ammonia produced by bacteria. pp. 133–136. in *Automation, Mechanisation and Data Handling in Microbiology*, (A. Baillie & R. J. Gilbert, eds). Society for Applied Bacteriology Technical Series No. 4. Academic Press: London and N.Y.

BUCK, P. W., ELSWORTH, R., MILLER, G. A., SARGEANT, K., STANDLEY, J. L. & WADE, H. E. (1971). The batch production of L-asparaginase from *Erwinia carotovora. J. gen. Microbiol.*, **65,** i.

CAPIZZI, R. L., BERTINO, J. R. & HANDSCHUMACHER, R. E. (1970). L-Asparaginase. *Ann. Rev. Med.*, **21,** 433.

COONEY, D. A. & HANDSCHUMACHER, R. E. (1970). L-Asparaginase and L-asparagine metabolism. *Ann. Rev. Pharmacol.*, **10,** 421.

COWAN, S. T. & STEEL, K. J. (1965). *Manual for the Identification of Medical Bacteria.* Cambridge: University Press.

DEALY, J. D. & UMBREIT, W. W. (1965). The application of automated procedures for studying enzyme synthesis in *Escherichia coli. Ann. N.Y. Acad. Sci.*, **130,** 745.

EDWARDS, P. R. & EWING, W. H. (1972). *Identification of Enterobacteriaceae.* 3rd edn., Minneapolis: Burgess Publishing Company.

GERKE, J. R. & FERRARI, A. (1968). Review of chemical and microbiological assay of antibiotics. pp. 531–541, in *Automation in Analytical Chemistry*. Technicon Symposia (1967) Vol. I. Mediad, White Plains, N.Y.

GERKE, J. R., HANEY, T. A., PAGANO, J. F. & FERRARI, A. (1960). Automation of the microbiological assay of antibiotics with an autoanalyzer instrument system. *Ann. N.Y. Acad. Sci.*, **87,,** 782.

KOCH, F. C. & MCMEEKIN, T. L. (1924). A new direct Nesslerization micro-Kjeldahl method and a modification of the Nessler-Folin reagent for ammonia. *J. Am. Chem. Soc.*, **46,** 2066.

LECLERC, H. (1967). Mise en évidence de la décarboxylase de l'acide glutamique chez les bactéries a l'aide d'une technique automatique. *Ann. Inst. Pasteur Lille*, **112**, 713.

LOWRY, O. H., ROSENBROUGH, N. J., FARR, A. L. & RANDALL, R. J. (1951). Protein measurement with the Folin-phenol reagent. *J. biol. Chem.*, **193**, 265.

POLLARD, A. & WALDRON, C. B. (1967). Automatic radioimmunoassay. pp. 49–59, in *Automation in Analytical Chemistry*. Technicon Symposia (1966) Vol. I. Mediad, White Plains, N.Y.

ROBERTS, J., BURSON, G. & HILL, J. M. (1968). New procedures for purification of L-asparaginase with high yield from *Escherichia coli. J. Bact.*, **95**, 2117.

TRINEL, P. A. & LECLERC, H. (1972). Automation de l'analyse bacteriologique de l'eau—I. Étude d'un nouveau test specifique de contamination fecale et des conditions optimales de sa mise en evidence. *Water Res.*, **6**, 1445.

WADE, H. C., ROBINSON, H. K. & PHILLIPS, B. W. (1971). Asparaginase and glutaminase activties of bacteria. *J. gen. Microbiol.*, **69**, 299.

WILLCOX, W. R., LAPAGE, S. P., BASCOMB, S. & CURTIS, M. A. (1973). Identification of bacteria by computer III. Statistics and programming. J. gen. Microbiol. **77**, 317.

An Automatic Plating-out Machine for Microbiological Assay Employing an 8×8 Design

D. A. SYKES AND C. J. EVANS

Pharmaceutical Production Control Laboratory,
Beecham Research Laboratories, Worthing, Sussex, England

Two of the primary considerations for a control laboratory, which is on routine assay of large numbers of samples, are precision of results and speed of throughput. Under these circumstances automation is essential in order to realize the maximum productive capacity of the staff and premises. Such automated techniques are gaining wider acceptance and some legislative bodies have now given approval for the use of automated assay methods employing the same microbiology as the official methods, provided the results obtained are of equivalent accuracy (Anon, 1972).

Much has been done to automate the continuous flow assay systems as reviewed by Kuzel and Kavanagh (1971 *a, b*) but rather less for large plate assays (Guartsveld, 1964; Di Cuollo, Guarini and Pagano, 1965; Levin, 1968; Clare 1973). All manual assay techniques depend heavily on the skill and care exercised by the technician and this is particularly true in the case of filling and reading large plates. The repetitive nature of the task is not conducive to the daily production of the high standard of work required and it is not surprising therefore that human errors occur. Reading errors tend to be insidious and can appear as biased results or a general loss of precision. In this laboratory, which is engaged in antibiotic assay, the difficulties of reading have been overcome by the use of an automatic, zone reading instrument (Grady and Sykes, 1970). This has been in routine operation for more than three years and the automation of this part of the assay procedure had the effect of emphasizing variations due to manual plating-out.

The instrument described here (Figs 1 and 2; Autodata, 80 Walsworth Road, Hitchin, Hertfordshire), was developed to overcome the problems associated with manual plating-out thereby improving assay precision. Parallel assays have shown the automated results to be more consistent

FIG. 1. Plating-out machine with access covers in the closed position. Note the punched tape reader and control switches in the left hand cover and the sample unit to the left of the instrument.

FIG. 2. Plating-out machine with access covers in the open position showing the metering pumps and an assay plate. The sample rack and buffer tank, partly withdrawn from the sample unit, are also shown.

than those achieved manually by a competent laboratory technician and variation in zone diameters from the same solution has been greatly reduced.

Ancillary equipment

The 12 in square assay plates used with the filler are the same as those described by Grady and Sykes (1970) for use with the plate reader. Substitution of toughened, 0·25 in plate glass for normal 0·25 glass has been found to give a much longer life to the plate. Although some sheets may have to be rejected initially, owing to distortion during the toughening process, the long term saving is considerable. It has also proved possible to dispense with the rubber gasket between the glass and the frame by increasing the number of retaining spring-clips from four to eight. Modified assay plates of this type are supplied by Autodata (80 Walsworth Road, Hitchin, Herts).

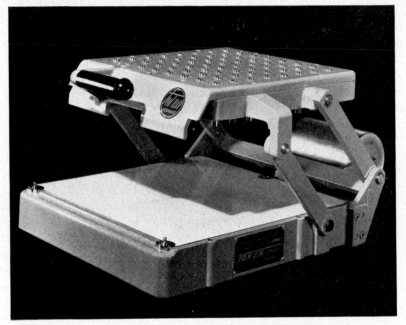

FIG. 3. Multiple punch showing locating pins.

Precise location of the plate is achieved in both the plate filler and reader by means of pins which engage in two locating blocks attached to one of the outer edges of the plate frame. Similar spring-loaded pins are fitted to the front edge of the plate punching device (Fig. 3). This consists of 64 individually spring-loaded punches arranged in an 8×8 array, and is capable of punching the 8 mm cups symmetrically at $1\frac{3}{16}$ in centre-distances. The use of this machine is essential in order to provide the precise location of the cups required for the filler and the reader.

The plating out machine

The instrument automates completely the task of filling the 64 cups in a 12 in square assay plate with antibiotic solutions. It is programmed by punched tapes that dictate the order in which the cups are filled, the duration of the rinse-cycles and co-ordinate the position of the assay plate in relation to the filling head. For a typical assay using one standard and three samples, all at two concentrations, a random Latin square design is used and eight cups are filled with each appropriate solution. However, a programme can be written for any assay design within the limits of eight solutions and 64 cups arranged in a 12 in square plate. Quasi-Latin square designs, e.g. 16 solutions each at 4 levels, may be filled by running the same plate twice through the machine, filling the first 32 cups with solutions 1–8 and the remaining cups with solutions 9–16.

There are eight completely separate liquid handling channels, one end of each is dipped automatically into either a container of rinsing fluid or a container holding one of the eight antibiotic solutions. The other end of each channel terminates in a mobile filling head above the surface of the assay plate. The plate and the filling head can be moved in relation to each other so that any channel-end may be brought above any of the 64 cups in the agar. The dose for each cup is delivered by means of a positive-displacement metering pump in each channel which has been designed to ensure high precision and drip-free filling in spite of the inevitable mechanical wear. The traction on each pump is maintained by means of a tensator and the rate of discharge is governed by a cantelever and mass system (vibrating wire and weight). To ensure uniformity of dose and to avoid splashing or drop formation, it is particularly important that the movement of the plunger should be constant. The metering pumps are activated by individual solenoids (Fig. 4). Each operation of a solenoid allows the plunger of a metering pump to move forward a distance sufficient to displace one dose. This distance is determined by the tooth-to-tooth pitch of an escapement rack fitted to each unit. Each cup is filled with 80 μl $\pm$ 5·0% and the filling speed is 2 min per plate which includes the rinsing operations. Allowing a short time for the loading of plates and sample solutions, a single technician can achieve a rate of about 15–20 plates/h.

All parts of the instrument which come into contact with liquid are resistant to corrosion by buffer or antibiotic solutions at the concentrations at which they are used. Manual flushing, additional to that provided by the automatic sequence, can be produced by operating the level shown to the left of the bank of metering pumps in Fig. 4. Although the principal liquid handling parts are readily removable for cleaning or maintenance, "in place" flushing with 0·1% (w/v) "Pyroneg" (Diversey Ltd., Cockfosters, Barnet, Herts) has been found to maintain the machine in a clean condition.

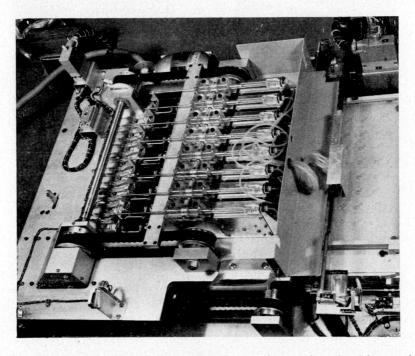

FIG. 4. Main covers removed from the plating-out machine to show the eight metering pumps. Note also the bail which re-sets the pumps supported by the rubber driving belts and the manual flushing lever situated at the extreme left of the instrument.

Operation of the filler

An assay plate is inserted into the carriage on the right hand side of the instrument (Fig. 2) and locked into position by means of the catch shown in Fig. 5, this ensures that the plate is situated correctly on the locating pins. The mechanism is interlocked so that the instrument cannot be started unless this catch is closed properly. The hinged access covers are

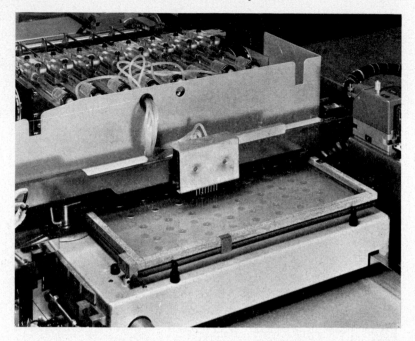

FIG. 5. Main covers removed to show filling head over partly filled assay plate. Note locking catch on left of plate.

interlocked so that the machine will not function unless they are also closed.

The paper-tape programme of choice is inserted into the tape reader situated in the left hand cover of the instrument (Fig. 1). The 8 solutions are placed in disposable medicine measures (Sterilin Ltd., 12–14 Hill Rise, Richmond Surrey) and these are introduced into the sampling unit by means of a removable rack. Figure 6 shows the rack in position in the sampling unit (cover removed). The instrument cannot be started unless the sample rack is inserted correctly. Additional sampling racks can be used in order to facilitate rapid loading of the solutions. At this point the sample probes are immersed in a reservoir containing buffer solution which is kept full by means of a constant level device.

From the operation of the start-button, the whole sequence is automatic. As the programme begins the assay plate is moved from right to left and buffer solution is pumped through the system by means of the 8 channel peristaltic pump shown in Fig. 7. During this part of the programme the filling head is situated over a drain and the rinse fluid is run to waste. The buffer rinse lasts approximately 11 sec and experiment has

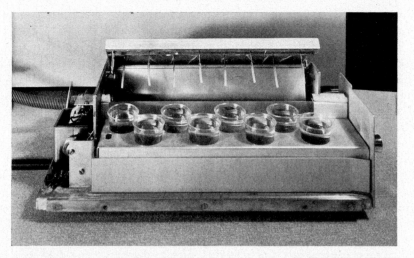

FIG. 6. Sampling unit with covers removed showing sample rack in position. The sample probes are mid-way between the buffer tank and sample cups. The actuating mechanism can be seen at the left of the unit.

FIG. 7. Eight-channel peristaltic pump housed under the bank of metering pumps. This section can be raised into the vertical position for access.

shown this to be sufficient to remove all detectable antibiotic from the system.

The next part of the programme commences when the peristaltic pump stops for approximately 1 sec while the sample probes move from the buffer reservoir to the sample cups. Figure 6 shows the probes midway between the buffer tank and the sample cups. The movement is achieved by means of a solenoid and linkage, part of which can be seen on the left in Fig. 6. The peristaltic pump re-starts and flushes each channel with the corresponding antibiotic solution thereby ensuring that all the metering pumps are fully charged and all the buffer solution has been displaced. This part of the programme lasts a further 11 sec and, during this time, the assay plate reaches the full extent of its travel so that the first row of cups are in line with the filling head. The peristaltic pump has now stopped and the bail, which pulls the plungers back at the end of the programme, has moved to the right to the position shown in Fig. 4.

Each metering pump is activated in turn to discharge one dose to waste which removes any droplets left hanging on the delivering tubes by the rinsing procedure. This movement also closes the inlet by which fluid enters the pump. The filling head then moves across the plate and locates the appropriate delivery tube, as determined by the punched tape programme, over the first cup. The corresponding metering pump is activated and an 80 μl dose is discharged into the cup. As the programme proceeds, the filling head moves along the row positioning the required tube end over the appropriate cup until the row has been filled. The plate moves in eight steps, each time bringing a fresh row of cups beneath the filling head. When all the cups have been filled the plate returns to the starting point while the bail moves the plungers of the metering pumps back to their original positions. The plate retaining-catch can then be released and the plate removed for incubation. Fresh sample solutions and a different Latin square programme can be inserted if required and, after introducing another assay plate, the instrument is ready to re-commence the filling operation.

The authors wish to express their thanks to the management of Beecham Research Laboratories for permission to publish this paper.

References

ANON. (1972). *Tests and methods of assay of antibiotic and antibiotic-containing drugs* **37**. Federal Register, No. 74, 7498.

CLARE, A. R. (1973). Specialised microbiological methods. In *Automated Analysis of Drugs and other Substances of Pharmaceutical Interest*. (Chapter 7), Butterworth: London.

Di Cuollo, J. C., Guarini, J. R. & Pagano, J. F. (1965). Automation of large plate agar microbiological diffusion assays. *Ann. N.Y. Acad. Sci.*, **130**, 672.

Grady, A. E. & Sykes, D. A. (1970). Automated reading of large assay plates in automation, mechanisation and data handling. p. 77. In *Microbiology*. A. Baillie and R. J. Gilbert (Eds). Academic Press: London.

Guartsveld, F. U. G. (1964). Mass detection of antibiotics in milk. *Tijolsche Diergeneesk*, **89**, suppl. II.

Kuzel, N. R. & Kavanagh, F. W. (1971*a*). Automated system for analytical microbiology I: Theory and design considerations. *J. Pharm. Sci.*, **60**, 764.

Kuzel, N. R. & Kavanagh, F. W. (1971*b*). Automated system for analytical microbiology II: Construction of system and evaluation of antibiotics and vitamins. *J. Pharm. Sci.*, **60**, 767.

Levin, J. D. (1968). Application of partial automation to agar-plate microbiological assays. *Ann. N.Y. Acad. Sci.*, **153**, 595.

Virus Concentration by Means of
Soluble Ultrafilters

S. F. B. POYNTER, H. H. JONES* AND J. S. SLADE

Thames Water Authority, New River Head Laboratories, 177 Rosebery Avenue, London EC1R 4TP, England

At present the hygienic quality of water is assessed chiefly by chemical and bacteriological standards. However surveillance of water for the presence of viral pathogens is now regarded as necessary because of conclusive evidence for the spread of viral infections *via* water supplies and swimming pools (Berg, 1967). Furthermore, the presence of viral pathogens has led to a review of the procedures for decontamination of water because treatments designed for the removal of bacteria may not be adequate for the removal of viral pathogens (Clark *et al.*, 1962). Effective monitoring of viral contaminants surviving decontamination procedures requires sensitive methods for their isolation. Several methods have been tried in laboratory and field studies including ultracentrifugation (Anderson *et al.*, 1967), aqueous two-phase separation (Shuval *et al.*, 1967) and adsorption on to insoluble polyelectrolyte (Wallis *et al.*, 1969). One of the most promising methods has been filtration through alginate membranes, first developed by Schyma (1960) from studies into carbohydrate-gel membranes by Thiele and Schyma (1953). The use of these membranes for isolation of small viruses was first discussed by Gärtner and Schnubrien (1964), and Witt (1964). Later Gärtner (1967), Poynter (1970) and Nupen (1970) reported recoveries of between 25 and 100% of the input virus. However no detailed studies of the conditions for optimal functioning of these membranes have been published which may account in part for the fact that although they are quite efficient, other methods are still being sought (Hill *et al.*, 1971) for the isolation of small numbers of virus from potable water supplies.

At the Metropolitan Water Board, alginate ultrafilters (Sartorius) have been used for a number of years (since 1964) for examination of sewage effluent and water samples from rivers, storage reservoirs and filtered

* Present address: The Department of Biological Sciences, University of Surrey, Guildford, England.

66 S. F. B. POYNTER *et al.*

chlorinated waters passing into supply and it is therefore timely to provide an assessment of the method and details of its operation. The commercial membranes are composed of a mixture of lanthanum and aluminium alginates stabilized with glycerol (Gärtner, 1967). They are in-depth type filters capable of retaining viruses and are supplied sterile on cellulose ester support membranes. For use they are placed in a standard filter holder and positive or negative pressures can be applied. A rare and attractive feature is their solubility in isotonic sodium citrate (3·8% w/v) giving a solution which is non-toxic to tissue cultures.

Materials and Methods*

Filtration equipment

This consisted of a stainless steel pressure vessel (Sartorius), stainless steel, gamma filter units (Whatman) and Gallenkamp membrane filter apparatus (47 mm). The pressure vessel was sterilized by chlorination and the remaining equipment by autoclaving at 121°/15 min.

Virus

The virus used was an attenuated poliovirus I (Strain Lsc 2ab). Stock virus for filtration experiments was prepared in VERO cell cultures (Yasamura and Kawakita, 1963) by inoculating three-day-old confluent cultures with 10^6 TCD_{50} virus. The virus was harvested after 3–4 d incubation at 37° by subjecting the culture to three successive periods of rapid freezing and thawing. The cell suspension was clarified by filtration through a 0·22 μ membrane (Millipore GS), a 0·1 μ membrane (Millipore VC) and a 0·05 μ membrane (Millipore VM). The object of the final membrane was to minimize the size and numbers of viral aggregates.

Cell culture

Growth medium. To Medium '199' (Burroughs Wellcome) was added foetal calf serum (5%v/v Tissue Culture Services), bicarbonate–carbon dioxide buffer (0·1% v/v), penicillin (200 units/ml) and streptomycin (100 units /ml).
Maintenance medium. This was the same as growth medium except that the percentage of foetal calf serum was halved to 2·5% and the bicarbon-ate-carbon dioxide buffer concentration doubled to 0·2%.
Overlay medium. This was maintenance medium gelled with 1·2% (w/v) agar (Difco purified). Intra-vital neutral red (0·004% w/v) was added

* The addresses of suppliers are given in the Appendix.

prior to overlaying. In addition to penicillin and streptomycin, the following antibiotics were added Neomycin sulphate (70 μg/ml), Polymyxin B (100 units/ml) and Fungizone (3·5 μg/ml).

Stock cultures were grown in 20 oz glass bottles in growth medium. The growth medium was replaced by maintenance medium when the cell sheet was confluent, usually at day 3–4. Cultures for plaque assay were grown in 4 oz bottles and used at an early stage of confluent growth, usually day 3–4, having been changed from growth medium to maintenance medium a day previously.

Plaque assay. Confluent cultures were drained, the maintenance medium discarded, and the monolayer inoculated with 0·5 ml of the test sample of virus suspension. After 1·5 h at room temperature 10 ml of agar overlay medium was added and the cultures immediately covered by a black cloth. When the overlay medium had solidified the cultures were incubated in the dark at 37° with the monolayer surface inverted. Plaque counts were made at day 3 and each subsequent day until the maximum number of plaques appeared.

Filtration and Ultrafiltration of the Sample

Filtration

To avoid clogging of the alginate membrane, ultrafiltration must be preceded in the case of turbid waters by filtration through a filter or filters of suitable porosity, the choice of which is dependent on the nature of the water or other liquid to be tested. Virus loss by adsorption at this stage must be avoided by treating the membrane before use by the passage of a dilute protein solution such as 5% foetal calf serum, broth or 1% bovine albumen. Alternatively the filter may be soaked in the protein solution. In practice it was found convenient to remove membranes as required from half-strength Hartley's digest broth kept sterile by simmering. It has been found that this treatment prevents any significant virus loss (see Table 1).

Relatively small samples of up to 3 l may be filtered through Oxoid cellulose acetate membranes (47 mm diam) or some equivalent bacterial grade of cellulose ester membrane. The numbers of cellulose ester membranes required will depend on the volume and turbidity of the sample and with very turbid samples prior filtration through a clarifying grade of membrane filter may be called for. In the case of River Thames water volumes of between 50 and 250 ml can be filtered through an Oxoid membrane before clogging necessitates the introduction of a new membrane.

TABLE 1. Virus recoveries before and after filtration through 'Oxoid' membrane treated with Hartley's Digest Broth

Water volume	PFU/ml before filtration	PFU/ml after filtration
Tap water, 1000 ml	79	71
Distilled water, 1000 ml	65	52
Tap water, 1000 ml	43	45

FIG. 1. Large volume filtration apparatus.

For large volumes of water (so far the largest volume tried is 20 l) Whatman gamma 30 filters were used. These again were treated by flushing with neat Hartley's digest broth prior to use. Two filter cartridges grade 30–80 (8 μ) and 30–03 (0·3 μ) were connected in line and the sample forced through under pressure at 5 lb in^2 (see Fig. 1). No loss of virus was detected during this filtration stage (see Table 2).

TABLE 2. Filtration of 15 litres of River Thames water through treated Whatman
gamma 30 filters

	Virus PFU/ml	Bact. count/ml at 22°
Before filtration	30	15,000
	26	12,000
After 1 litre	26	240
	40	
After 7 litres	36	200
	42	
After 14 litres	41	320
	28	

Filtrates derived from treating turbid waters can be ultrafiltered without
further treatment and from this point the same procedures are followed as
in the case of initially clean waters such as London tap water.

Ultrafiltration

The Gallenkamp filter apparatus may be used for ultrafiltration since the
Sartorius alginate membranes have the same diameter as those used in the
preceding stages. The base of the holder is connected via a waste water
reservoir to a source of vacuum. When a number of filtrations are necessary as many vacuum points as required can readily be provided by means
of a manifold (Fig. 2). The manufacturers claim that the alginate membranes are able to withstand negative pressures up to 700 mm mercury and
this was borne out in practice. The filtration rate depended on the turbidity and the degree of negative pressure applied and some typical
results are shown in Table 3.

TABLE 3. Negative pressure, turbidity and flow rate

Water	Negative pressure mmHg	Flow rate through alginate membrane ml/min
Distilled	100	1
	650	8
Tap	100	0·2
	650	2–3
River	650	1

When ultrafiltration was completed the membrane was removed and
dissolved and the receiver containing the filtrate was disinfected if necessary.

FIG. 2. Alginate ultrafiltration apparatus.

Solution of the alginate membrane

The alginate filter with its support membrane was placed in a sterile petri-dish. To this was added at least 2 ml (see below) of sterile 3·8% (w/v) sodium citrate. A pipette was used to irrigate the alginate to ensure solution before drawing off the liquid into a sterile receptacle. Storage at 30°C did not affect the titre in the case of poliovirus type 1.

Efficiency

The factors studied in the assessment of the alginate method were the effect of pH in the range 4·5–8·5, the temperature in the range 4–25°, negative pressure in the range 100–700 mmHg, the input level of virus and the volume of sodium citrate used to dissolve the membrane. The results are given in Table 4. Continuing studies are investigating the same factors using other types of water and other viruses namely Coxsackie-viruses A9 and B4, Echovirus 1 and Reovirus 1 and these results will be reported elsewhere.

It appears that recovery was affected by the volume of sodium citrate used, the optimal volume being between 4 and 10 ml. In practice 4 ml was normally used as a larger volume decreased the concentration factor and increased tissue culture requirements.

Recovery was also affected by the degree of negative pressure used, maximum recoveries being at 400 and 600 mmHg. There was no apparent effect of pH in the range 4·5–8 but above this pH a structural change took place in the membrane which allowed virus particles to pass through. Temperature and concentrations of input virus did not seem to affect recoveries in the ranges tested. Lower temperatures are recommended as this reduces the possibility of viral inactivation. In assessing the sensitivity of the alginate method the lowest level of virus tried was 8 PFU/litre and 4 PFU/litre were recovered indicating that the method is indeed very sensitive. It must be stressed here that these results are all based on attenuated poliovirus 1 and other viruses may show different profiles.

The reason for the loss of a large percentage of the virus has been closely examined by several methods:

(a) Examination of the filtrate after alginate filtration showed that in the majority of cases no virus could be detected and the maximum detected was 6% of the input virus.

(b) No significant virus levels (< 1%) could be detected by eluting the support membrane with foetal calf serum and no increase in titre in the filtrate was noted when the normal support membrane was replaced by a broth treated "Oxoid" membrane.

TABLE 4. Effect of certain factors on recovery of poliovirus 1 from alginate membranes

Test		Type of water	Initial virus concentration PFU/ml (average)	No. of replicate samples	Mean recovery expressed as % input virus	Standard deviation
Sodium citrate volume	0·5 ml	Distilled	50	5	14·1	8·9
	1 ml		50	5	21·6	8·0
	2 ml		50	5	30·6	6·9
	4 ml		50	5	40·6	9·1
	10 ml		50	2	44·0	—
	16 ml		50	2	20·0	—
Negative Pressure	100 mmHg	Distilled	100	4	32·5	5·6
	200 mmHg		68	4	42·0	11·0
	400 mmHg		40	4	70·0	6·1
	600 mmHg		72	4	50·0	3·5
pH	4·0	Distilled	52	3	40·5	7·7
	5·5		52	3	40·0	5·0
	7·0		52	3	42·5	11·7
	8·0		52	3	40·0	3·1
	8·5		52	3	6*	—
Temperature	4°	Distilled	66	3	45·0	4·1
	13°		66	3	46·0	12·7
	24°		66	3	42·0	12·9
Virus concentration	A	Distilled	16200	2	64·0	—
	B		1800	2	66·0	—
	C		180	2	70·0	—
	D†	River	8	2	50·0	—

*See text on pH. †Crude virus suspension.

(c) Treatment of the whole alginate membrane with Hartley's digest broth, which had been shown to prevent adsorption of viruses on to celulose ester membranes, had no effect on recovery rates confirming that alginate membranes are in-depth physical limiting membranes. Further evidence comes from electron micrographs published by Gärtner (1967) in which it can be seen that the membrane is in two layers and virus retention occurs at the interface.

Our observations show that under a fairly wide range of conditions the alginate membrane behaves in a moderately reproducible fashion. It was interesting to note that under the conditions tested we never recovered all the virus and although the loss of virus remains unexplained it was gratifying to note that the filtrate and support membrane remained relatively free of virus. This suggests that the virus was inactivated, formed an irreversible bond with the membrane components or perhaps formed stable aggregates which would result in decreased plaque forming activity. The evidence for improved yield dependent on volume of sodium citrate provides some support for the binding or aggregation phenomenon rather than inactivation. These results have been obtained with virus mono-dispersed by successive filtrations, whereas the use of crude virus suspensions such as exist in nature tend to give higher recoveries possibly due to disaggregation.

The method described has proved to be simple, effective and inexpensive and gives results which compare favourably with other methods.

Appendix

Suppliers of equipment

Burroughs Wellcome–Wellcome Reagents Ltd., Beckenham, Kent.
Difco Laboratories, P.O. Box 14B, Central Avenue, East Molesey, Surrey.
A. Gallenkamp & Co. Ltd., Technico House, 6 Christopher Street, London, EC2.
Millipore (U.K.) Ltd., Millipore House, Abbey Road, London, NW10.
Oxoid Ltd., 20 Southwark Bridge Road, London, SE1.
Sartorius–V. A. Howe & Co. Ltd., 88 Peterborough Road, London, SW6.
Tissue Culture Services Ltd., 10–12 Henry Road, Slough, Bucks.

References

ANDERSON, N. G., CLINE, G. B., HARRIS, W. W. & GREEN, J. G. (1967). Isolation of viral particles from large fluid volumes. In *Transmission of Viruess by the Water Route*, p. 75. G. Berg. (Ed.). Interscience Publ., London.

74   S. F. B. POYNTER *et al.*

BERG, G. (1967). *Transmission of Viruses by the Water Route.* Interscience Publ., London.

CLARK, N. A., BERG, G., KABLER, P. W., & CHANG S. L. (1962). Human enteric viruses in water; Source survival and removability. 523. *Proc. First Int. Conf. Wat. Poll. Res. London.*

GÄRTNER, H. & SCHNUBRIEN, R. (1964). *Arch. Hyg. Bakteriol.*, **148**, 183.

GÄRTNER, H. (1967). Retention and recovery of polioviruses on a soluble ultrafilter. In *Transmission of Viruses by the Water Route.* 121. G. Berg, (Ed.). Interscience Publ., London.

HILL, W. F., AKIN, E. W. & BENTON, W. H. (1971). Detection of viruses in water: A review of methods and applications. *Proc. 13th Water Quality Conf.* V. L. Snoeyink, (Ed.).

NUPEN, E. M. (1970). Virus studies on the Windhoek waste-water reclamation plant (South Africa). *Wat. Res.*, **4**.

POYNTER, S. F. B. (1970). Personal communication. In *Proc. 13th Water Quality Conf.* (1971). V. L. Snoeyink, (Ed.).

SHUVAL, H. I., CYMBALISTA, S., FATTAL, B. & GOLDBLUM, N. (1967). Concentration of enteric-viruses in water by hydro-extraction and two phase separation. In *Transmission of Viruses by the Water Route.* 45. G. Berg, (Ed.). Interscience Publ., London.

SCHYMA, D. (1960). *Zentr. Bacteriol. I Orig.*, **178**, 229.

THIELE, H. & SCHYMA, D. (1953). *Naturwissenschaften*, **40**, 583.

WALLIS, C., GRINSTEIN, S., MELNICK, J. L. & FIELDS, J. E. (1969). Concentration of viruses from sewage and excreta on insoluble polyelectrolytes. *Appl. Microbiol.*, **18**, 1097.

WITT, G. (1964). *Arch. Hyg. Bakteriol.*, Heft. 3, 188.

YASAMURA, Y. & KAWAKITA, Y. (1963). Research into SV40 by tissue culture. *Nippon Rinsho.*, **21**, 1201.

The Assay of Thermally Induced Inhibitors Derived from Nitrite

J. Ashworth, Linda L. Hargreaves, Anthea Rosser
and B. Jarvis

B.F.M.I.R.A., Randalls Road, Leatherhead, Surrey, England

For many years the safety and stability of canned cured meats have remained unexplained. Whilst the public health record of these products is excellent, it is noteworthy that the heat treatments (often only mild pasteurization) used for such products are insufficient to destroy spore-forming bacteria. Specifically these products have a remarkably good record of freedom from association with outbreaks of botulism. In a system as complex as a cured meat product, many factors must be involved, to a greater or lesser extent, in explaining this phenomenon. These include, in addition to the effect of the heat treatment *per se*, such things as the sodium chloride and sodium nitrite contents of the products, plus the initial low incidence of spore forming organism which are likely to be encountered in meat processed under hygienic conditions.

In 1967 it was shown that when a complex laboratory medium was heated with sodium nitrite, and subsequently challenged with an inoculum of vegetative cells of *Clostridium sporogenes*, the system was more inhibitory than was explicable by the sodium nitrite content *per se* (Perigo, Whiting and Bashford, 1967). Certain aspects of this inhibitory effect differed from those associated with sodium nitrite. Whilst the inhibitory activity of the latter was pH dependent, the inhibitory activity being ascribed to undissociated nitrous acid, that of the heated nitrite system was largely pH independent. At pH values around neutrality, levels of nitrite in excess of 100 ppm were needed to inhibit *Cl. sporogenes* when added to sterile medium, whereas 3–5 ppm nitrite heated in the medium were inhibitory. From these results Perigo *et al.* (1967) postulated that when nitrite was heated sufficiently in this complex medium, it was involved in a chemical reaction which yielded some unknown inhibitory substance. They suggested that this thermally induced inhibitor may play a role in the safety and stability of certain sub-lethally processed cured meats.

Although Labbe and Duncan (1970) confirmed the work of Perigo *et al.*

(1967) in a medium system, with *Clostridium welchii* as test organism, they were of the same opinion as Johnston, Pivnick and Samson (1969) in stating that a comparable effect did not occur in meat, i.e. a thermally induced inhibitor derived from nitrite was not produced in canned cured meats. Labbe and Duncan (1970) suggested that sodium chloride and sodium nitrite, combined with the heat treatment and low pH values, alone maintained the safety and stability of canned cured meats.

An effect, comparable in many aspects to that demonstrated by Perigo *et al.* (1967) in media, has been shown in a pork/nitrite system given a heat sterilization treatment (Ashworth and Spencer, 1972). Such a system bears little ressemblance to commercial conditions, but Ashworth, Hargreaves and Jarvis (1973) have observed the production of a thermally induced inhibitor in pork given a heat treatment (pasteurization) in the presence of nitrite at levels similar to those used commercially.

All this work demonstrates clearly the production during heating of one or more inhibitory compounds in certain meat and media systems containing nitrite. Because of the possible role of such inhibitors in meat cured commercially and also because of their considerable antimicrobial activity, further investigation of these and similar systems was desirable. The pursuit of such investigations, especially when considering the attempts to isolate and identify the inhibitor(s) produced by nitrite heated in meat and media systems, is dependent upon the availability of reliable, quantitative assay methods. The various assay techniques described here have been used in the investigation of the inhibitory medium/nitrite systems, and other inhibitory systems.

Inhibitors in a medium system

Tube bioassay techniques

In preliminary experiments undertaken in an attempt to confirm and elucidate further the observation of Perigo *et al.* (1967), the original bioassay described in their paper was used. This original assay was based upon the complex laboratory medium (Table 1). This was dispensed (20 ml) in McCartney bottles, and heated with various levels of nitrite, or alternatively heated alone and inoculated aseptically with a sterile solution of nitrite at appropriate levels. All systems were assayed by challenging them with a culture vegetative cells of *Cl. sporogenes*, followed by incubation at 30° for 5 d. The use of this bioassay procedure presented certain problems. A fairly large volume of sample (20 ml as a minimum) was needed and a long incubation period required. Moreover, there was poor reproducibility between experiments and a lack of consistency and sensitivity within any one experiment. Attempts were therefore made to develop alternative assay procedures.

TABLE 1. The complex laboratory medium* used to demonstrate the production of thermally induced inhibitors derived from nitrite

Tryptone (Oxoid)	20	g
Peptone (Oxoid)	10	g
Lab-lemco (Oxoid)	10	g
Yeast extract (Oxoid)	5	g
Sodium chloride AR (Fisons)	5	g
Dipotassium hydrogen phosphate AR (Fisons)	2·5	g
Glucose AR (Fisons)	2	g
Soluble starch AR (Fisons)	1	g
Sodium thioglycollate (BDH)	1	g
0·4% (w/v) alcoholic solution of bromocresol purple	5	ml
Sodium hydroxide or hydrochloric acid to pH 7·4	q.s.	
Distilled water to	1	l

* From Perigo, Whiting and Bashford (1967).

One possible explanation of the variability in results of the original bioassay was the development of aerobic conditions in the assay medium. Such conditions might either inhibit the growth of the clostridia in some bottles in which they might otherwise have grown, or might destroy the inhibitor if it were sensitive to oxidation, thus allowing growth of the clostridia in some bottles in which the organism might otherwise be inhibited. The addition of agar to the medium or the sealing of the surface of the medium with liquid paraffin B.P., or 2% (w/v) agar have been found to improve reproducibility. All replicates at one dilution of the test sample behave similarly and there is the expected proportional response. If these replicates are dispensed in test tubes (15 × 1·5 cm) rather than the Mc-Cartney bottles as used by Perigo et al. (1967) there is a marked increase in reproducibility possibly because the former aid in the maintenance of an anaerobic environment. By using 5 × 0·5 cm test tubes it was possible to reduce considerably the volume of sample required for the assay. Finally it has been found that the temperature of incubation can be increased from 30° to 37°, and at the latter temperature the incubation period can be reduced to as little as 1 d. As with the original procedure, however, there is a tendency for some tubes which do not initially permit growth to do so on prolonged incubation.

The application of these assays falls into two categories. If a medium is heated with nitrite, the production of a thermally induced inhibitor in the medium can be monitored by inoculation of suitable dilutions (in media) of the inhibitory systems with Cl. sporogenes (PA 3679). Alternatively, the assays may be applied to inhibitors extracted by suitable chemical means (e.g. water immiscible solvents) from media systems, or to any other inhibitory material, e.g. iron nitrosyl complexes such as Roussin salts.

Production and assay of inhibitors in media

The basic medium (Table 1) is dispensed (50 or 100 ml quantities or quantities suitable for dilution) into 5 oz round bottles, or (200 ml) into 10 oz medical flats. Up to 500 ppm sodium nitrite are added and a standard heat treatment of 115° for 20 min in a domestic pressure cooker is given. After cooling and dilution the medium is assayed in sterilized medium which may or may not contain agar. Alternatively, a small tube bioassay can be used. In this case 2·5 ml of the medium (Table 1) are dispensed into bijoux bottles. Sodium nitrite is then added by syringe in ≯ 25 μl amounts (up to 60 ppm final concentration), and the systems given the standard heat treatment. In either case the use of agar should be avoided if thioglycollate is present, because of the likelihood of additional inhibitor formation when agar, nitrite and thioglycollate are heated together.

The medium is inoculated at room temperature, or, if it contains agar, at 50°, with an 18 h Robertson's cooked meat broth (CMB) culture of the test organism, *Cl. sporogenes* (PA 3679). Fifty ml of media are inoculated with 5 drops (from a Pasteur pipette) of the culture, and 2·5 ml of the media are inoculated with 1 drop of the culture, representing inocula of about 15×10^6 and 3×10^6 organisms respectively.

After thorough mixing, the inoculated medium is dispensed into 4 test tubes, of either $15 \times 1·5$ cm in the case of the large, or of $5 \times 0·5$ cm in the case of the small bioassay. When no agar has been incorporated into the medium, the surface of the liquid in each tube is sealed with a layer of either liquid paraffin or, preferably, molten agar (2% w/v). All tubes are incubated for a minimum of 24 h at 37°, and examined for growth. This is assessed by acid production if a seal of agar is used, gas production is also evident in the liquid assay.

Assay of inhibitory extracts

Solvent extracts from systems of media heated with nitrite, and also other inhibitory complexes, e.g. Roussin salts, do not contain the necessary factors for the growth of the test organism, and therefore appropriate nutrients must be added. The samples are sterilized in 40 or 2 ml amounts, either by heat or by filtration. Where heat is used for sterilization, agar (e.g. Oxoid No. 3; 1·2% w/v) can be added before the treatment. Where sterilization by filtration is used for the thermolabile samples, the liquid bioassay is used. After sterilization of the test samples, basal medium containing $\times 5$ the normal concentration of solids (subsequently referred to as concentrated medium) is added; 10 ml are added to 40 ml of the test solution, and 0·5 ml to 2 ml of the test solution. The samples are then bioassayed in a manner similar to that described above.

Comparison of the various liquid bioassays

The original type of assay used by Perigo et al. (1967) has been compared with the modified tube assay incorporating agar (at a level of 1·2% w/v). In both cases the medium was heated with various levels of nitrite, and then these solutions were challenged with Cl. sporogenes, and dispensed in small volumes (10 ml) into semi-bijoux bottles, and test tubes respectively. Whilst it is possible that the agar heated with the medium and nitrite may, itself, have given rise to an inhibitor, examination of the results obtained with these two systems showed similarities in the overall response. With the agar assay, however, the response was basically "all or none" (although it was possible to grade partially the degree of growth), whereas a proportional response was obtained with the original assay (i.e. there was a tendency for a variation in response between replicates derived from the same sample).

The medium was also heated with 500 ppm nitrite and, after cooling, a range of dilutions of this sytem were prepared using pre-sterilized medium both with and without agar. After inoculation with Cl. sporogenes, the dilutions were dispensed in either semi-bijoux bottles or test tubes and all were incubated at 37°. In this case, because the agar was not heated with nitrite together with the medium, there was no complication in interpretation of results due to the production of an additional inhibitor.

The "all or none" response obtained from the agar tube assay is clearly seen in the results (Table 2), and this contrasts with the proportional response obtained using the original type of liquid assay. A marked variation in the response occurred in the original assay system, which can be minimized by use of the agar systems. The latter also has the advantage of allowing early interpretation of results, i.e. after one day.

Both the large and small modified liquid tube assays were compared using the basic medium (Table 1) and also an acid hydrolyzed casein (AHC) medium. The AHC medium is a modification of the basic medium, in which the protein hydrolysate components of the medium, i.e. peptone, tryptone, Lab-lemco and yeast extract, are replaced by 6% (w/v) AHC. Because of the high salt content of the AHC (35% w/w sodium chloride), no additional salt is included in the medium. In both cases sodium nitrite at levels in the range 0–60 ppm were added to the media before the heat

Table 2. Comparison of the original liquid and the agar bioassays using dilution

Dilution	Assay method			
	Agar, days at 37°		Original, days at 37°	
	1	2	1	2
1:5	0/40*	0/40	0/50	12/50
1:10	40/40	40/40	7/50	32/50
1:20	40/40	40/40	39/50	42/50
1:50	40/40	40/40	42/50	45/50

* Results are expressed as the number of cultures showing growth out of the total number of replicates tested

treatment. After this treatment and inoculation with *Cl. sporogenes*, the media were dispensed into either 15×1.5 cm or 5×0.5 cm tubes, and the surface of the liquid in each tube sealed with liquid paraffin. The results (Table 3) show that the actual volume of the assay system has little bearing upon the results, but that there is a difference in the concentration of nitrite required to produce an inhibitory effect when heated in the two media.

Optimum inoculum size

To obtain reasonable sensitivity from the bioassay, the size of inoculum used is important. A typical overnight (18 h) culture of *Cl. sporogenes* in CMB has been found to contain *c.* 8×10^7 organisms/ml. Inocula of between 3×10^4 and 15×10^6 organisms/ml from an 18 h culture of *Cl. sporogenes* in CMB were investigated in the test system. The inoculum size was found to have a definite effect on the inhibitory activity during incubation, and has, therefore, been standardized at one drop (from a Pasteur pipette) of an 18 h CMB culture ($\sim 3 \times 10^6$ organisms) to 2.5 ml medium and five drops ($\sim 15 \times 10^6$ organisms) to 50 ml medium.

Agar diffusion bioassay

With continual use, several disadvantages became apparent with the various modified bioassays described above. The results obtained from these assays were mainly qualitative. This prevented the application of the bioassay for the quantitative assay of thermally induced inhibitors. Whilst the modified bioassays were reasonably reproducible, they still needed a fair sized sample (minimum of 2 ml), were rather insensitive, and time consuming to do. They were subject also to interference by residual nitrite, sodium chloride, and several solvents which were used in the preparation of some of the samples, thus no single bioassay procedure was applicable to all samples.

A new bioassay system based upon the punched plate agar-diffusion technique has therefore been developed. This system is most suitable for aerobic organisms, but can also be used with anaerobes. Quantitative assays against an internal standard are feasible, and only a very small sample is required (less than 0.25 ml). The assay is insensitive to many solvents, as well as to relatively high levels of sodium chloride and sodium nitrite. This one assay can be applied to all samples derived from medium systems, but unfortunately, no comparable punched plate system has yet been developed for inhibitory meat systems.

The antimicrobial spectrum of a thermally induced inhibitor from the

TABLE 3. Comparison of large and small scale liquid bioassays

Input Sodium nitrite level (ppm)	Incubation at 37° (d)							
	1 day				2 days			
	Basic medium*		AHC medium**		Basic medium		AHC medium	
	Small scale	Large scale	Small scale	Large scale	Small scale	Large scale	Small scale	Large scale
0	+***	+	+	+	+	+	+	+
6	+	+	+	+	+	+	+	+
12	+	+	+	+	+	+	+	+
24	+	−	−	−	+	+	−	−
36	−	−	−	−	+	+	−	−
48	−	−	−	−	+	+	−	−
60	−	−	−	−	+	+	−	−

* See Table 1; ** details on page 79, and *** results are based on the response of four replicates for each treatment; +, all replicates grew, and −, no growth.

medium/nitrite system has been examined by several workers. Perigo and Roberts (1968) demonstrated that the inhibitor was effective against a range of clostridia in addition to *Cl. sporogenes*, and Labbe and Duncan (1970) demonstrated the sensitivity of *Cl. welchii* to such an inhibitor. Roberts and Garcia (1973) have shown that the thermally induced inhibitor in the medium/nitrite system is inhibitory against a range of bacilli and also *Streptococcus durans*. Ashworth *et al.* (1974) have shown that the inhibitor produced in this system of nitrite heated with the medium is effective against a wide range of microorganisms, including both Gram positive and Gram negative aerobes. In this work both *Staphylococcus aureus* and *Staph. saprophyticus* were shown to have a sensitivity to the inhibitor similar to that of *Cl. sporogenes*.

Experimental details of the agar diffusion bioassay

Molten nutrient agar (Oxoid) at $\sim 50°$ is inoculated with an 18 h culture of *Staph. saprophyticus* in nutrient broth to give a final concentration of $\sim 10^6$ organisms/ml. The seeded agar (23 ml) is then dispensed aseptically

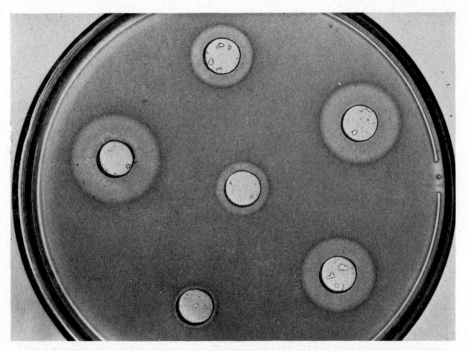

FIG. 1. Agar diffusion bioassay of ammonium heptanitrosyl trithiotetraferrate (Roussin black salt). Test organism, *Staphylococcus saprophyticus*.

in sterile polystyrene Petri dishes (Sterilin, 85 mm diam), resting on a level bench. When the agar has set and cooled, wells of 7·0 mm diam are punched in the agar. Fixed volumes (0·1 ml) of various dilutions of the sample under test, and of set concentrations of a standard, are introduced into the wells in the plates.

For a standard, the inorganic iron nitrosyl co-ordination complex, Roussin black salt (ammonium heptanitrosyl trithiotetraferrate) is used. Experimental examination has shown that the antimicrobial spectrum of this compound (Dobry and Boyer, 1944; Candeli and Mancini, 1948) is very similar to that of the inhibitor produced in a heated medium/nitrite system (Ashworth et al., 1974).

The plates are incubated overnight at 37° and the diameters of the zones of inhibition measured with vernier callipers. By comparison with the standard, a quantitative estimation of the relative potency of samples derived from medium systems is possible (the relative dose/response of Roussin black salt and these samples is similar). Where it is necessary to use Cl. sporogenes, molten Reinforced Clostridial Agar (Oxoid) at 50° is inoculated with ∼ 1 × 10⁶ organisms/ml and the plates incubated at 37° in anaerobic jars.

TABLE 4. Comparison of agar diffusion (punched plate) bioassay and small tube bioassay of ammonium Roussin black salt

	Assay method	
Roussin black salt (ppm)	Agar diffusion Staph. saprophyticus (Mean* diameter (mm) of zone of inhibition	Small tube Cl. sporogenes (No. of replicates (4) showing growth)
0	—	4/4
2·5	7·9	4/4
5	9·38	0/4
10	10·96	0/4
25	13·53	0/4
50	15·18	0/4
100	16·95	0/4

* Mean of four diameters of duplicate zones.

The following example (Fig. 1) indicates clearly the quantitative nature of the results obtained with this new bioassay using ammonium Roussin black salt. For comparison, the results obtained using the modified small tube bioassay (liquid with a plug of agar) are given in Table 4. The results obtained for the agar diffusion bioassay may be represented graphically (Fig. 2) and thus provide a bioassay standard. The results from a bioassay of an inhibitory extract from acid hydrolyzed casein (AHC) medium heated with nitrite are also included in this graph.

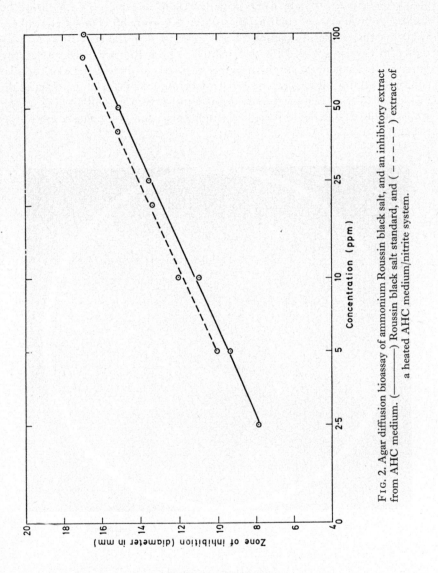

FIG. 2. Agar diffusion bioassay of ammonium Roussin black salt, and an inhibitory extract from AHC medium. (————) Roussin black salt standard, and (– – – – –) extract of a heated AHC medium/nitrite system.

It is interesting to note that some surface active agents cause inactivation of thermally induced inhibitors from the medium system, and also of ammonium Roussin black salt. This effect is only apparent with surface active agents with high hydrophile/lipophile balance, e.g. hydrophilic surface active agents such as Brij 35 and Tween 80 (Honeywill-Atlas). Figure 3 shows the interaction of a 0·005% w/v solution of Brij 35 (contained in the central well), with 50 and 100 ppm Roussin black salt in the outer wells. A zone of interaction has occurred at the interface of the two solutions in the agar, resulting in distortion of the zone of inhibition produced in the lawn of *Staph. saprophyticus*. Subsequently some of the Roussin black salt has diffused through this zone and caused inhibition beyond.

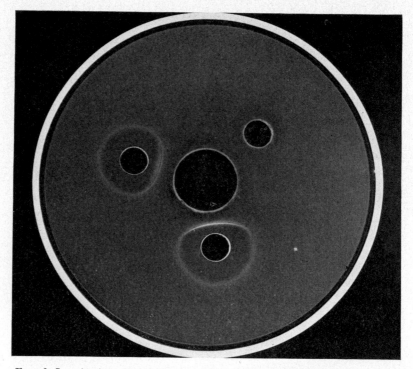

FIG. 3. Inactivation of the inhibitory activity of Roussin black salt, by Brij 35.

Inhibitors in meat

The original assay used by Perigo *et al.* (1967) has been adapted for application to meat systems. This meat assay, fully described by Ashworth

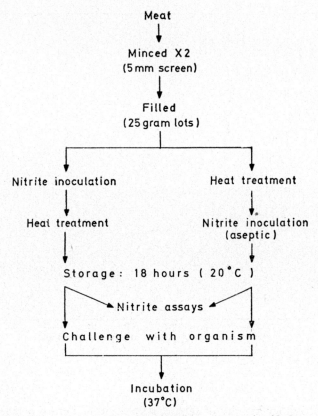

F IG. 4. Protocol of assay of thermally induced inhibitors in meat (pork). Organisms tested: *Clostridium sporogenes*, *Clostridium botulinum* (types A, B, E). Normal heat treatment: 115° for 20 min.

and Spencer (1972), can be diagrammatically represented as in Fig. 4. Twenty five g of minced pork (or 22·5 g minced beef) are filled into wide mouthed 1 oz Universal bottles with serum caps. Half the bottles are heated with various levels of nitrite, whilst the remainder are heated alone. When the latter are cold, sterile sodium nitrite, at various levels, is added aseptically. Five replicates at each nitrite level are set up, and after sufficient time has elapsed to ensure even distribution of the nitrite throughout the meat plug, four replicates at each nitrite level are challenged with $\sim 4 \times 10^6$ vegetative cells of the test organism, *Cl. sporogenes* PA 3679 (or *Cl botulinum* types A, B or E). At the time of challenge the residual nitrite levels are determined on the remaining non-inoculated bottle at each nitrite level, so as to take into account the nitrite which is lost on heating or during storage of the system. After incubation for various periods at

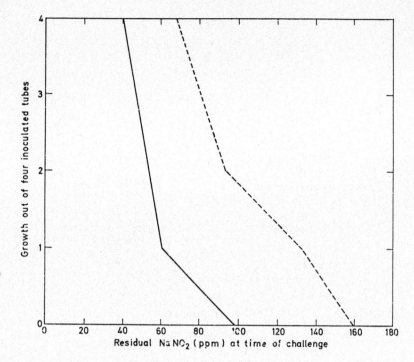

FIG. 5. Production of a thermally induced inhibitor, derived from nitrite, in pork. (————). Nitrite added before heat, and (– – – – –) nitrite added after heat.

37°, the inhibitory activity of nitrite, added either before or after heating is determined. A typical set of results is illustrated in Fig. 5. From these results it is possible to determine the ED_{50} values of each of the two treatments employed, i.e. those levels of residual nitrite in the meat (at the time of challenge with the test organism), required to cause inhibition in 50% of the samples, based on the quantal growth response at the various nitrite levels (Ashworth and Spencer, 1972). In this particular case, the ED_{50} values at 5 d are 54 ppm and 94 ppm nitrite, when the nitrite was added, either before or after the heat treatment respectively. This demonstrates that when nitrite is added before heating the resultant inhibition cannot be explained by the residual nitrite *per se* but must also result from some other factor(s) produced during the heat treatment. This assay is also applicable to beef, and the results obtained in an experiment with lean shin beef are represented graphically in Fig. 6.

The use of this assay, for the identification of thermally induced inhibitors produced in systems of meat heated with nitrite, has several drawbacks, not least of these is that the assay is very time consuming. It

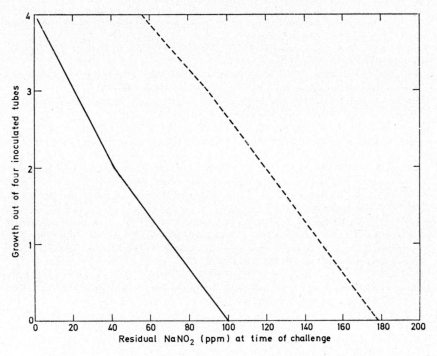

FIG. 6. Production of a thermally inhibitor, derived from nitrite in beef. (————————)
Nitrite added before heat, and (– – – – –) nitrite added after heat.

is therefore unfortunate, that attempts to adapt the agar diffusion plate bioassay for the assay of thermally induced inhibitors in meat, have, to date, proved singularly unsuccessful.

Acknowledgements

The authors wish to express their gratitude to all their colleagues at the Research Association who have freely given help and advice at various stages in this work, and particularly to Miss Merete M. Giles for her considerable contribution in the development of the modified liquid and agar tube assays.

References

ASHWORTH, J., HARGREAVES, L. L. & JARVIS, B. (1973). Production of an antimicrobial effect in pork heated with sodium nitrite under simulated commercial pasteurization conditions. *F. Fd Technol.*, **8**, 477.
ASHWORTH, J., DIDCOCK, A., HARGREAVES, L. L., JARVIS, B., LARKWORTHY,

L. F., & WALTERS, C. L. (1974). Chemical and microbiological comparisons of inhibitors derived thermally from nitrite with an iron thionitrosyl (Roussin black salt). *J. gen. Micn.*, **84** (in press).

ASHWORTH, J. & SPENCER, R. (1972). The Perigo effect in Pork. *J. Fd Technol.*, **7.** 111.

CANDELI, A. & MANCINI, M. (1948). Bacteriostatic effect of potassium ferro-heptanitrosotrisulphide. *Boll. Soc. med-chir. Modena.*, **48**, 161.

DOBRY, A. & BOYER, F. (1944). On the antiseptic action of iron nitrosulphide or Roussin's salt. *Anals. Inst. Pasteur, Paris*, **71**, 455.

JOHNSTON, M. A., PIVNICK, H. & SAMSON, J. M. (1969). Inhibition of *Clostridium botulinum* by sodium nitrite in a bacteriological medium and in meat. *Can. Inst. Fd Technol. J.*, **2**, 52.

LABBE, R. G. & DUNCAN, C. L. (1970). Growth from spores of *Clostridium perfringens* in the presence of sodium nitrite. *Appl. Microbiol.*, **19**, 353.

PERIGO, J. A. & ROBERTS, T. A. (1968). Inhibition of clostridia by nitrite. *J. Fd Technol.*, **3**, 91.

PERIGO, J. A., WHITING, E. & BASHFORD, T. E. (1967). Observations on the inhibition of vegetative cells of *Clostridium sporogenes* by nitrite, which has been autoclaved in a laboratory medium, discussed in the context of sublethally processed cured meats. *J. Fd Technol.*, **2**, 377.

ROBERTS, T. A., & GARCIA, C. E. (1973). A note on the resistance of *Bacillus* spp., faecal streptococci and *Salmonella typhimurium* to an inhibitor of *Clostridium* spp. formed by heating sodium nitrite. *J. Fd Technol.*, **8**, 463.

The Assay of Nisin in Foods

G. G. FOWLER, B. JARVIS AND J. TRAMER

Aplin & Barret Ltd., Yeovil, Somerset.
British Food Manufacturing Industries Research Association,
Randalls Road, Leatherhead, Surrey.
Unigate Ltd., Central Laboratory, Western Avenue, London W3, England

Nisin is the name given to a group of large polypeptide antibiotics (dimeric particle weight, 6800), produced by strains of *Streptococcus lactis*; it occurs naturally in many cheeses. In many countries nisin may be added to foods as a preservative. Nisin inhibits the outgrowth of germinated bacterial spores and is particularly effective against gas-producing clostridia and thermophilic sporeformers which might spoil canned foods and processed cheese. Its use as a preservative has been tested also in meat products such as meat paste, pasteurized ham, tongue and luncheon meat, although its use for these products is restricted at the present time (Tramer, 1966; Jarvis and Morisetti, 1969).

In the U.K. nisin is permitted without limitation on amount and without the need to declare its use on the label in processed cheese, clotted cream and canned foods (Anon, 1962). In this context, canned foods are defined as being foods in hermetically sealed containers which have been sufficiently heat processed to destroy any spores of *Clostridium botulinum* in food or container, or which have a pH value of less than 4·5.

Although there is no legal quantitative limit to the use of nisin in foods, it is essential for control purposes to have methods for its detection and estimation. It is desirable, also, to have a means of identification in order to adjudicate, if necessary, on the nature of any materials detected in foods by a bioassay procedure. This is particularly important in instances where spurious materials, which are purported to be nisin, might be used as food preservatives.

Methods for the Estimation and Identification of Nisin

Many bioassay procedures have been used from time to time for the estimation of nisin, including tube dilution techniques, turbidimetric assays, agar diffusion assays and estimates of bactericidal or sporicidal effect. The tube dilution method of Friedman and Epstein (1951) is very rapid, giving results within some 30–60 min, and is based upon the use of redox indicators for endpoint determination with cultures of *Streptococcus cremoris* in milk. However, it is insufficiently accurate for control purposes. Estimation of the inhibition of lactic streptococci by measurement of final pH value and of the bactericidal effect of nisin (Hirsch, 1950) are too inaccurate and cumbersome for routine use. The turbidimetric assay of Berridge and Barrett (1952), and the modification of this method by Hurst (1966), is rapid, sensitive and permits a good throughput of samples. However, the opalescence of many food extracts precludes the use of this technique for the estimation of nisin in foods although it is an excellent research tool. A novel method, proposed by Stumbo, Voris, Skaggs and Heinemann (1964), measures the relative inhibition by nisin of the recovery of heated spores of *B. stearothermophilus*. This method is very sensitive but is too involved and time-consuming for routine application and requires many replicates for good reproducibility.

A number of agar diffusion procedures have been proposed with a variety of test organisms including strains of *Lactobacillus bulgaricus*, *Str. cremoris*, *Bacillus megaterium*, *B. stearothermophilus* (spores), and *Micrococcus flavus* (Beach, 1952; Mocquot and Léfèbre, 1956; Tramer and Fowler, 1964). In the U.K., the preferred methods are those described originally by Tramer and Fowler (1964). One of the two methods reported by these workers is a modification of the semi-quantitative reverse phase assay developed for penicillin in milk by Kosikowski and Ledford (1960). In the nisin assay, spores of *B. stearothermophilus* are suspended in a non-nutrient agar; germination and outgrowth from the spores will not occur until assay discs containing nutrients are dipped into a food extract and applied to the surface of the agar. Diffusion of nutrients stimulates germination and outgrowth under suitable conditions of incubation and results in zones of exhibition around the assay discs. When the food extract contains an inhibitor such as nisin, the assay disc is surrounded by a zone of inhibition which is itself surrounded by a zone of exhibition (Fig. 1). An advantage of this method is that the plates may be prepared in bulk for despatch from a central laboratory to control laboratories in the factory; the spores remain viable for some weeks provided that the plates are sealed in plastic bags and stored under refrigeration.

The other method of Tramer and Fowler (1964) uses *M. flavus* as test organism and the medium contains Tween 20 to aid diffusion of nisin through the agar. Acid extracts of the food material under test are distributed in wells punched into the medium and the presence of antibiotic is shown by the development of zones of inhibition. Compensation for the presence of interfering substances extracted from the food is made by the use of specially prepared diluents in which both test and standard solutions are prepared. The extraction and assay procedure given below is based on modifications to the original method resulting from discussions of a British Standards Institute committee. The extraction procedure has been modified to permit the introduction of tests for heat stability under extremes of pH value and for assay also against a nisin-producing strain of *Str. lactis*. These procedures permit the differentiation of nisin from many other antibiotics and food preservatives which are used in, or which might adventitiously contaminate, food materials.

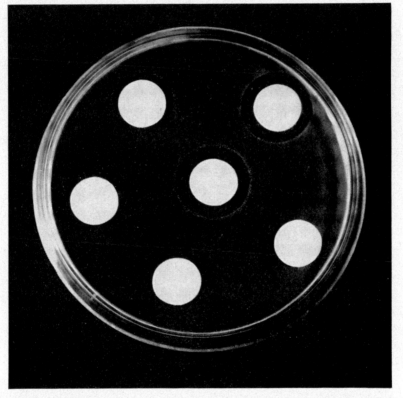

FIG. 1. Reverse phase assay plate showing zones of inhibition produced by different concentrations of nisin.

A method of differentiation for materials which purport to be nisin is based upon the use of a nisinase (dehydroalanine reductase) produced by some strains of *Bacillus cereus* and related organisms (Jarvis, 1967; Jarvis and Berridge, 1969; Jarvis and Farr, 1971). An active preparation of the enzyme is reacted in buffer against unknown and standard preparations of nisin followed by assay of initial and residual antimicrobial activity.

The Semi-Quantitative Reverse Phase Assay

Materials

Assay medium

Contains (g l^{-1}): sodium chloride, 9, and Agar No. 1 (Oxoid), 15.

Sporulation agar

Contains (g l^{-1}): proteose peptone (Oxoid), 20; sodium chloride, 5, and Agar No. 1 (Oxoid), 15.

Nutrient discs

Whatman AA discs (13 mm) dipped for 5 sec in sterile 20% (w/v) peptone solution and dried *in vacuo* overnight over phosphorus pentoxide.

Spores

An actively growing culture of *B. stearothermophilus* (NCIB 8224) is inoculated into Glucose Nutrient broth and incubated for 16 h at 55°. The cells are harvested by centrifugation, washed twice with sterile distilled water and re-suspended in a further volume of sterile distilled water. The suspension is spread on the surface of the sporulation agar contained in Petri dishes, Roux bottles or large "medical flat" bottles, which are incubated at 55° until extensive (> 90%) sporulation has occurred (usually 2–3 d). The spores are washed from the surface of the agar with sterile distilled water and the washings are collected. The suspension of spores is stored at 4° until required for use.

Preparation of the assay plates

The suspension of spores is heat shocked in a boiling water bath for 10 min and is then inoculated into the saline non-nutrient agar to give a final concentration of c. 4×10^5 spores ml^{-1}. Ten ml quantities are poured into levelled, sterile flat-bottomed Petri dishes and allowed to solidify. If the plates are sealed into polythene bags they can be stored for some time at 4°, but they should not be stored at room temperature otherwise a rapid reduction in the number of viable spores will occur.

Food extract

A 20% (w/v) suspension of the food is prepared by emulsifying 20 g food with *c*. 60 ml 0·02 N HCl, adjusting the pH to 2·0 with 5 N HCl and heating for 5 min in a boiling water bath. The extract is cooled to room temperature, adjusted to pH 6·5 with 5 N NaOH and diluted to 100 ml with distilled water.

Procedure

A nutrient-assay disc is immersed in the pH-adjusted test solution, touched against a filter paper to remove excess solution and placed on the surface of the agar. When all test and standard solutions have been applied, the assay plate is incubated for 18–24 h at 55°. The zones of inhibition produced within the zones of exhibition are measured and compared with the zones produced by standard preparations to give a semi-quantitative estimate of nisin concentration.

Discussion

The diameter of the zone of inhibition produced is dependent not only upon the nisin concentration but also upon the amount of interfering substances present in the food extract. To some extent compensation for this can be made by the use of nisin-free diluents (see below) for the preparation of standard solutions. However, normally a concentration of 0·5–1·0 i.u. ml^{-1} is required to produce a measurable zone of inhibition and by the use of suitable dilutions a semi-quantitative estimate of the nisin concentration of a food can be obtained without recourse to comparison with standard preparations. This makes the method eminently suitable for routine control purposes especially where laboratory facilities or suitably skilled personnel are not available for more sophisticated assays. For some foods, the extraction process with acid is not necessary and the test can be undertaken directly upon the food extract—this applies particularly to vegetable and milk preparations containing nisin. However, with meat products, acid extraction is essential to release the nisin, but it is necessary to neutralize the acid before undertaking the assay since the test organism is sensitive even to the effects of dilute acid used alone.

The Quantitative Agar Diffusion Assay used for the Estimation and Differentiation of Nisin in Foods

Materials

Assay Medium

The assay medium contains (g l^{-1}) Bacteriological peptone (Oxoid), 10; Lab-lemco (Oxoid), 3; sodium chloride, 3; yeast extract (Marmite), 1·5; natural, soft, raw cane sugar from Barbados (Lion Brand Barbados Muscovado) 1, and Agar No. 1 (Oxoid) 10. The pH of the medium is adjusted to 7·5 ± 0·1 prior to sterilization for 20 min at 121°.

Test Organisms

Micrococcus flavus (NCIB 8166) is maintained by culture on slopes of the assay medium incubated at 30° for 48 h. *Streptococcus lactis* (NCIB 8586) is maintained on slopes of the assay medium which are incubated at 30° for 24 h. Prepared slopes of both organisms may be stored at 4–7°, until required, up to a maximum of 14 d.

Nisin Standard

A secondary standard nisin, calibrated against the international reference preparation, is available from Aplin & Barrett Ltd., Yeovil, Somerset. The standard preparation contains 1000 i.u. mg^{-1} dry weight.

Reagents

The following reagents are required: 0·02 N HCl; 5 N HCl; 5 N NaOH. These solutions should be prepared from Analar grade reagents and should not be made up from ampoules of concentrated analytical reagent which may contain traces of materials inhibitory to the test organism.

Apparatus

In addition to normal laboratory apparatus, a centrifuge with a capacity of not less than 200 ml and a rotor speed of at least 670 **g** is required, together with a water bath maintained at 63 ± 1°. Assay plates should be flat bottomed with minimum internal dimensions of not less than 270 × 270 × 6 mm. A suitable levelling table and spirit level are also required.

Preparation of the assay plates

The inoculum of test organisms is prepared by emulsifying the growth from a slope culture with 10 ml of quarter-strength Ringer solution. The optical extinction of the suspension of cells is adjusted to give a transmission of 50% (equivalent to an optical extinction of 0·3) in a 10 mm cuvette at 650 nm. To the melted and tempered (50 ± 2°) assay medium is added 2% (v/v) of a 1 + 1 dilution of Tween 20 (polyxoyethylene sorbitan monolaureate; Honeywell-Stein Ltd., Carshalton) which has been held for 20–30 min at 50 ± 2°. The medium is mixed thoroughly and 2% (v/v) of the inoculum is added to the medium. After mixing well, the medium is left for a few minutes in the tempering bath, to permit dispersion of the foam, and the medium is then poured into sterile, previously levelled assay plates to give a depth of 3–4 mm agar. After the agar has set, the plates are inverted and stored at 4–7° for 1 h to facilitate the cutting of wells.

Wells are punched in the agar; using a cork borer or other suitable punch of 7–9 mm diameter, allowing at least 30 mm between adjacent wells and between peripheral wells and the edge of the assay plate. The spacing and number of wells will be dependent to some extent upon the nature of the assay design used. The discs of agar are removed from the plate using a sterile mounted needle or by means of a suitable vacuum device.

Preparation of food extracts

Forty ± 0·1 g of the food material, is weighed into a 500 ml beaker or the bag for a Colworth Stomacher (A. J. Sewell & Co. Ltd., London) and dispersed evenly in 160 ml 0·2 N HCl. Whilst for foods such as processed cheese, dispersion can be achieved by adequate stirring with a glass rod, the use of the "stomacher" aids dispersion for many food materials. Using 5 N HCl the pH is adjusted to 2·0 ± 0·1 and the slurry is then heated in a boiling water bath, such that the extract achieves a temperature of not less than 98° for 5 min. The extract is cooled rapidly to 20 ± 5° and the volume is adjusted to 200 ml with 0·02 N HCl. The dispersion is centrifuged for at least 10 min at 670 g and is then transferred to a refrigerator at 4–7° until the fat phase has solidified. The supernatant liquid below the fat later is decanted and filtered through glass wool. This is *Extract A*, of which 20 ml is required for assay, 20 ml for preparation of *Extract AX* and 60 ml for preparation of *Extract B*.

Twenty ml of *Extract A* is heated in a boiling water bath for a further period of 5 min. It is then cooled to 20 ± 5° and adjusted to the original volume using distilled water. This is *Extract AX*.

To prepare *Extract B*, 60 ml of *Extract A* is adjusted to pH 11·0 ± 0·3 with 5 N NaOH and is heated for 30 min at 63 ± 1°. The extract is cooled to 20 ± 5° and reacidified with 5 N HCl to pH 2·0 ± 0·1. Twenty ml of this extract is required for assay. To a further 40 ml of *Extract B* is added

120 ml of 0·02 N HCl. This is *Extract C* which is used as a diluent for both test and standard preparations.

Preparation of standard solutions

The standard nisin preparation (100 $\pm$ 1 mg) is dispersed in *c*. 80 ml of 0·02 N HCl and heated in a boiling water bath to not less than 98° for 5 min. It is cooled rapidly to 20 $\pm$ 5°, allowed to stand for 2 h, then diluted to 100 ml with 0·02 N HCl. This solution contains 1,000 i.u. nisin ml^{-1} and is stable for 7 d at 4–7°. For assay, it is diluted to the required levels in *Extract C*. At least three levels (10, 5 and 2·5 i.u. ml^{-1}) should be prepared; it is essential to prepare these each day as required and not to store them.

Estimation and differentiation of nisin

Three dilutions each of *Extract A, AX and B* are prepared in the diluent (*Extract C*). For most purposes dilutions of 1 + 3, 1 + 7 and 1 + 15 are suitable. With a calibrated Pasteur or dropping pipette equal volumes ($\not< 0$·10 ml) of standard and test solutions are delivered into wells punched into the agar. An appropriate method should be used for randomization and replication of the samples (Humphrey and Lightbown, 1952; Simpson, 1963). The plate is incubated for 18–20 h at 30 $\pm$ 1°, and the diameters of the zones of inhibition are measured along at least two diameters using an appropriate projection device or by means of needle point calipers.

Interpretation of the results

With dilutions of *Extract A* and of the standard nisin preparation a straight line should be obtained when the log nisin concentration is plotted against the zone diameter (Fig. 2). The potency of the sample can be calculated by comparison with the graph of the standard preparations, or by a suitable statistical method (Finney, 1952; Humphrey and Lightbown, 1952; Simpson, 1963).

Since nisin is stable to heat at pH 2·0, identical zone diameters (within practical limits) should be obtained from the tests on *Extract A and AX*, providing that nisin is the only inhibitor present in the extract. No zones of inhibition should be obtained using *Extract B* since nisin is destroyed rapidly when heated at pH 11·0.

Differentiation of nisin by assay against Streptococcus lactis

The assay plates are prepared as described above, except that an inoculum of *Str. lactis* is used in place of the *M. flavus*. For this test Petri dishes may be used instead of large assay plates. Dilutions of *Extract A* and of the standard nisin solution, prepared as described above, are placed in wells punched in the agar. After incubation for 18–20 h at 30°, no zones of inhibition should be seen if the test material contains only nisin.

Discussion

The method described above has been shown over many years of practical evaluation to be suitable for the routine assay of nisin in foods. The degree of precision and reproducibility of the assay will be dependent upon many factors. Where a high degree of precision is required, the food extract and standard solution can be assayed using an appropriate latin or

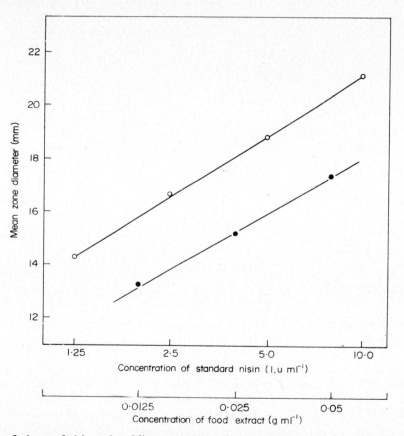

FIG. 2. Assay of nisin against *Micrococcus flavus*. Example of the graph obtained when the mean zone diameters are plotted against the $\log_{10}$ of the concentration of nisin in the standard solution (○) or against the $\log_{10}$ of the concentration of food extract (●).

quasilatin design and the concentration of nisin may be derived from the appropriate statistical calculation (Finney, 1952; Simpson, 1963). Although precision of this nature may be required for research purposes it is rarely necessary for routine application and the concentration of nisin can be obtained from a graph prepared from the diameters of the zones produced by standard solutions of nisin tested on the same assay plate (Fig. 2). It is essential to prepare standard solutions in the appropriate diluent made from the initial extract of the food under test, otherwise spurious results may be obtained. This was demonstrated by Tramer and Fowler (1964) who showed that enhanced diffusion of nisin occurred in the presence of emulsifying salts such as polyphosphates. Some of the latter are themselves capable of causing small zones of inhibition on the assay plate (Jarvis and Quinn, unpublished data). Fig. 3 illustrates the effect of preparing standard

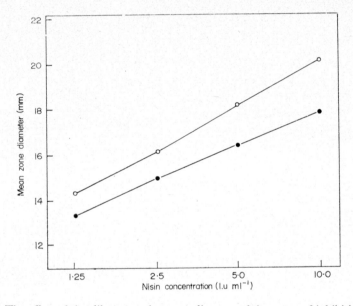

FIG. 3. The effect of the diluent on the mean diameter of the zones of inhibition in the *Micrococcus flavus* assay of nisin. The standard solution of nisin was diluted in acid extract C of a food material (○) or in 0·02 NHCl (●).

solutions of nisin in dilute acid and in the acidified alkali-treated diluent prepared from a food product. Since the slopes of the two lines diverge, it is obviously essential for the standard to be prepared in the same diluent as the test solution, otherwise an invalid assay result is obtained. The acid stability of nisin is of obvious value in permitting extraction of nisin from the food material without excessive extraction of other food con-

TABLE 1. Differentiation of nisin from other antimicrobials

	Stable to heat and acid; unstable to heat and alkali; or both	Inhibitory to *Streptococcus lactis*
Nisin	yes	no
Subtilin	yes	no
Polymyxin	yes	no
Tylosin	yes	yes
Gramicidin	no	no
Bacitracin	no	yes
Penicillins	no	yes
Chloramphenicol	no	yes
Erythromycin	no	yes
Neomycin	no	yes
Novobiocin	no	yes
Streptomycin	no	yes
Tetracycline	no	yes
Coliston	no	
Sorbic Acid	no	
Potassium Sorbate	no	
Sodium Nitrate	no	
Benzoic Acid	no	
Calcium Propionate	no	
Hydrogen Peroxide	no	
Sodium Citrate	no	
KH$_2$PO$_4$	no	

stituents. However, because of the adsorption of nisin to the proteins in food it is essential to use an acid extraction procedure in order to maximize recovery. Emulsification of the food without acid treatment leads to very low recovery of nisin in most instances.

The combination of stability to heat in acid solution and the instability to heat under conditions of alkaline pH value are of considerable value in differentiating between nisin and other antimicrobial agents which might be present in food materials. Table 1 lists those antimicrobial agents from which nisin can be differentiated on the basis of this test. The antibiotics polymyxin, subtilin and tylosin behave similarly to nisin but give lines of different slope. As mentioned previously, polyphosphate preparations used as emulsifying agents produce small zones of inhibition but the size of the zone does not vary significantly between different dilutions of the food extract and can be compensated for in the preparation of standard solutions. Formaldehyde and hexamethylene-tetramine which may be present in cheese extracts may produce zones of inhibition against the test organisms. Of the antibiotics listed in Table 1 all except nisin, subtilin polymyxin and gramicidin give zones of inhibition against the nisin-producing strain of *Str. lactis*.

On the basis of these tests it is possible to differentiate nisin from all antimicrobials listed with the exception of polymyxin and subtilin. Polymyxin can be differentiated, if necessary, by its inhibition of Gram negative bacteria and subtilin is not available commercially.

The "Nisinase" Method for Identification of Nisin

Enzyme preparation

This is prepared (Jarvis and Farr, 1971) by extraction of freeze-dried sporulated cells of *B. cereus* (NCDO 1937) using 0·05 M citric acid and 0·1 M Na_2HPO_4 buffer. The enzyme preparation may be partially purified or used as a crude extract.

Standard nisin solution

A standard solution is prepared to contain 5000 i.u. nisin ml^{-1} 0·02 N HCl as described above, except that 500 mg $\pm$ 5 mg nisin is dissolved in 100 ml 0·02 N HCl.

Test solutions

Solutions of test materials are prepared to give a nisin concentration equivalent to 5000 i.u. nisin ml^{-1} when assayed against *M. flavus* using the method described above.

Preparation of incubation mixtures

Into each of a series of test tubes, pipette 6·8 ml of citrate-phosphate buffer (pH 7·0; McIlvaine, 1921), 0·1 ml 10 mM cobalt nitrate solution and 0·1 ml 10 mM calcium chloride solution. Pipette 1 ml of the standard nisin solution into 2 tubes and 1 ml of each test solution into 2 tubes. To one tube in each pair add 1 ml of enzyme preparation which has been heated for 5 min in a boiling water bath and to the other tube add 1 ml of active enzyme preparation. Mix and incubate for 6 h at 30° $\pm$ 1°. At the end of the incubation period, add 1 ml 2N HCl to each tube, mix by inversion and heat the tubes in a boiling water bath for 5 min. Cool to 20 $\pm$ 5°, adjust the volume if necessary to 10 ml using 0·02 N HCl and prepare dilutions of the test and standard solutions in 0·02 N HCl.

Assay of residual nisin

Assay residual nisin using the *M. flavus* assay described above. It is preferable to use latin or quasi-latin square distribution and statistical analysis of results to obtain the degree of precision required. Using this method the overall reproducibility of the enzyme assay is about $\pm 8\%$.

Interpretation of results

An example of the results which have been obtained in one comparison of unknown materials is given in Table 2. The activity of the nisinase preparation is confirmed from the inactivation of the standard nisin preparation by the enzyme. A significantly lower residual nisin activity should

TABLE 2. Example of the results obtained when nisinase reacted against nisin standard and unknown preparations

	Mean residual activity (i.u. ml^{-1})* after incubation with:		Conclusion
	Nisinase	Heat-inactivated nisinase	
Standard nisin preparation	92	455**	Enzyme active
Unknown sample A	58	421	Nisin present
Unknown sample B	485	490	Nisin not present

* Determined by comparison with a standard solution of nisin. ** The theoretical nisin activity of 500 i.u. ml^{-1} incubation mixture will not normally be obtained since some (c.10%) chemical inactivation occurs during the incubation.

be seen in all cases where active enzyme has been incubated with a solution of nisin. Enzymic activity has not been observed against the antibiotics gramicidin, bacitracin, polymyxin, chlor- or oxytetracycline, benzylpenicillin or tylosin, nor against any antimicrobial agent which is commonly used as a food preservative (Jarvis, 1967; Jarvis and Farr, 1971).

Discussion

The nisinase of *B. cereus* has been shown to inactivate nisin and subtilin, two antibiotics which are dependent for antibiotic activity on the terminal amino acid sequence, dehydroalanyllysine. Since subtilin is not commercially available and has a different UV spectrum, these antibiotics may be readily differentiated from one another. Nisins A & B are the major constituents of commercial nisin which also may contain smaller quantities of nisins C, D & E. Nisins A, B, C & E are all inactivated by the nisinase but nisin D, which has a different chemical composition (Berridge, Newton and Abraham, 1952), is not inactivated. The enzyme assay will never therefore go to completion since some activity will be associated with the small amount (*c.* 1–4%) of nisin D in the mixture. The assay has been used on one occasion to demonstrate that a particular fraction of commercial nisin was not nisin A as had been claimed. It has been used also to demonstrate that an antibiotic isolated from a strain of *Str. lactis* did not contain a significant amount of nisins A, B, C or E.

Since nisin is somewhat labile at neutral pH, total recovery of the input nisin is never achieved. Normally about 10% chemical inactivation is observed. It had at one time been anticipated that the enzyme might be of potential value for the identification of nisin in food extracts (Jarvis and Berridge, 1969). Unfortunately, at the concentration of nisin in foods and in the presence of certain unidentified components of foods, the activity of the enzyme is insufficient to obtain clear cut results. Whether the nisinase produced by strains of *Streptococcus thermophilus* (Alifax and Chevalier, 1962) might be more successful has never been investigated fully although preliminary investigations failed to reveal any significant advantages.

References

ALIFAX, R. & CHEVALIER, R. (1962). Study of the nisinase produced by *Streptococcus thermophilus*. *J. Dairy Res.*, **29**, 233.

ANON. (1974). *Preservatives in Food Regulations*. Statutory Instrument No. 1119: H.M.S.O., London.

BEACH, A. S. (1952). An agar diffusion method for the assay of nisin. *J. gen. Microbiol.*, **6**, 60.

BERRIDGE, N. J. & BARRETT, J. (1952). A rapid method for the turbidimetric assay of antibiotics. *J. gen. Microbiol.*, **6**, 14.

BERRIDGE, N. J., NEWTON, G. G. F. & ABRAHAM, E. P. (1952). Purification and nature of the antibiotic nisin. *Biochem. J.*, **52**, 529.

FINNEY, D. J. (1952). *Statistical Method in Biological Assay*. Charles Griffin, Ltd., London.

FRIEDMAN, R. & EPSTEIN, C. (1951). The assay of the antibiotic nisin by means of a reductase (resazurin) test. *J. gen. Microbiol.*, **5**, 830.

HIRSCH, A. (1950). The assay of the antibiotic nisin. *J. gen. Microbiol.*, **4**, 70.

HUMPHREY, J. H. & LIGHTBOWN, J. W. (1952). A general theory for plate assay of antibiotics with some practical applications. *J. gen. Microbiol.*, **7**, 129.

HURST, A. (1966). Biosynthesis of the antibiotic nisin by whole *Streptococcus lactis* organisms. *J. gen. Microbiol.*, **44**, 209.

JARVIS, B. (1967). Resistance to nisin and production of nisin-inactivating enzymes by several species of *Bacillus*. *J. gen. Microbiol.*, **47**, 33.

JARVIS, B. & BERRIDGE, N. J. (1969). The application of enzymes which inactivate antibiotics: a review. *Chemy. Ind.*, (1969), 1721.

JARVIS, B. & FARR, J. (1971). Partial purification, specificity and mechanism of action of the nisin-inactivating enzyme from *Bacillus cereus*. *Biochim. biophys. Acta*, **227**, 232.

JARVIS, B. & MORISETTI, M. D. (1969). The use of antibiotics in food preservation. *Int. Biodetn Bull.*, **5**, 39.

KOSIKOWSKI, F. W. & LEDFORD, R. A. (1960). A reverse-phase assay test for antibiotics in milk. *J. Amer. vet. Med. Ass.*, **136**, 297.

McILVAINE, T. C. (1921). A buffer solution for colorimetric comparison. *J. biol. Chem.*, **49**, 183.

MOCQUOT, G. & LÉFÈBRE, E. (1956). A simple procedure to detect nisin in cheese. *J. appl. Bact.*, **19**, 322.

SIMPSON, J. S. (1963). Microbiological assay using large plate methods. p. 129 in *Analytical Microbiology*, ed. by F. Kavanagh. Academic Press, London

STUMBO, C. R., VORIS, L., SKAGGS, B. G. & HEINEMANN, B. (1964). A procedure for assaying residual nisin in food products. *J. Fd. Sci.*, **29**, 859.

TRAMER, J. (1966). Nisin in food preservation. *Chemy. Ind.*, (1966), 446.

TRAMER, J. & FOWLER, G. G. (1964). Estimation of nisin in foods. *J. Sci. Fd. Agric.*, **15**, 522.

ADDENDUM

Since submission of this paper a revised British Standard has been published (BS 4020: February 1974) "Methods for the estimation and differentiation of resin in processed cheese".

Serological Methods for the Assay of Staphylococcal Enterotoxins

R. HOLBROOK AND A. C. BAIRD-PARKER

Unilever Research Laboratory, Colworth House, Sharnbrook, Bedford, England

Staphylococcal enterotoxins, which are the toxins responsible for staphylococcal food poisoning, are exotoxins elaborated by certain strains of *Staphylococcus aureus* growing in suitable environments (Baird-Parker 1971). They are single unbranched polypeptide chains with molecular weights of about 30,000 to 35,000. They are water soluble and highly resistant to physical and chemical treatments. (Bergdoll, 1967; Bergdoll, 1970; Bergdoll, 1972; Minor and Marth, 1972).

The proportion of *S. aureus* strains forming enterotoxins is in doubt. Casman *et al.* (1967) investigated the incidence of enterotoxin A, B, C and D producers amongst *S. aureus* strains isolated from various sources. Using serological tests they found that 44% of 438 clinical specimens, 31% of 144 isolates from nasal specimens from healthy individuals, 10% of 236 isolates from raw milk, 2% of 51 isolates from mastitic cows, 30% of 260 isolates from frozen foods and 96% of 80 isolates from food poisoning episodes were positive. Enterotoxins A and D alone or in combination occurred most frequently in all categories. However, about 30% of a random selection of the serologically negative strains were shown to produce cat emetic substances. In 1969, Bergdoll suggested in a personal communication, that about 50% of strains were enterotoxinogenic (Lachica, Weiss, and Deibel, 1969), but more recent work, using monkey feeding studies, showed that the incidence in strains from at least some environments appears to be much higher than this (Bergdoll, 1972). Hájek and Maršálek (1973) made a study of the incidence of enterotoxin A, B and C producers among *S. aureus* isolated from man, hens, swine, cows, sheep, hares, dogs, horses, mink and pigeons. *S. aureus* from these sources were classified into six biotypes. Over 99% of isolates from humans belonged to biotype A, and all but one of 491 isolates falling into the biotypes B to F were of animal origin. However strains of biotype A were isolated from domestic animals; 75% of 143 isolates of *S. aureus* of the A biotype from humans, and 28% of 47 isolates of *S. aureus* of the A biotype from domestic animals were enterotoxinogenic, but only 1·2% of 490 isolates of *S. aureus* belonging to the B to F biotypes isolated from animals produced enterotoxins A, B or C. It is therefore possible to conclude from this that the majority of *S. aureus* causing food poisoning are of the A biotype but not necessarily of direct human origin.

The enterotoxin yields by enterotoxinogenic strains of *S. aureus* varies considerably; some strains produce over 500 times as much toxin as others when grown under conditions optimal for enterotoxin synthesis. Strains which yield low levels of enterotoxin in laboratory media may produce sufficient enterotoxin when growing in a food to cause food poisoning; Bergdoll (1972) describes such an incident.

Purification of the enterotoxins from culture supernatants and immunization of rabbits with the purified toxin have enabled serological methods to be used for their detection and identification. So far five serologically distinct enterotoxins have been described in detail, they have been designated A to E (Casman, Bergdoll and Robinson, 1963). At least one further type, type F, is under active study by the FDA in the United States. Reviews of the purification procedures and chemistry of the enterotoxins are given by Bergdoll (1967, 1970).

The following methods and procedures for detecting enterotoxins have been developed over many years study in this and other laboratories. They can be recommended as reliable and suitable methods for use in a routine quality control laboratory.

Detection of Enterotoxinogenic Strains of *Staphylococcus aureus*

Many culture methods have been devised for detecting enterotoxin production by *S. aureus*. These methods include dialysis sac culture techniques (Casman and Bennett, 1963; Donnelly, Leslie, Black and Lewis, 1967), growth on cellophane overlaid on agar (Hollander, 1965; Jarvis and Lawrence, 1970), growth in semi-solid agar and growth in shaking liquid cultures (Kato *et al.*, 1966; Reiser and Weiss, 1969; Dietrich, Watson and Silverman, 1972).

The two sac culture methods tend to give higher concentrations of toxin (μg/ml) than the other methods and using these methods we often achieve a 100 fold increase in toxin concentration over that obtained in a stationary broth culture of the same medium. Bergdoll (1972) considers the method of Casman and Bennett (1963) to be superior because the toxin is contained in a smaller final volume. However Untermann (1972) recommends the Donnelly method, whilst Šimkovičová and Gilbert (1971) found little difference in toxin concentration between the two methods and, in agreement with these workers, we have found the Donnelly method easier and quicker to perform, on a routine basis.

Brain Heart Infusion Broth is most frequently used as the growth medium for enterotoxin detection, but different commercial brands and different batches of the same brand vary in their ability to support good enterotoxin production (Casman and Bennett, 1963; Reiser and Weiss, 1969). It is therefore important to determine that a batch of medium supports enterotoxin production using known low toxin producing strains before it is used to determine the enterotoxinogenicity of unknown strains.

Sac Culture Method (Donnelley et al., 1967)

Dialysis tubing (24/32 Visking Tubing, Scientific Intrument Centre, 1, Leeke St., London W.C.1.) is boiled in water containing 0·01% (w/v) E.D.T.A. for 5 min and then washed in several changes of distilled water. A 60 cm length of tubing is tied tightly with two knots at one end and filled with water to check for leaks. Eighty ml of double-strength Brain Heart Infusion Broth (Fisher Scientific Co., Fair Lawn, New Jersey, U.S.A.) adjusted to pH 6·8 is poured into the sac. The air is expelled from above the liquid and the mouth tied with two further knots so that a loosely filled sac is obtained. The sac is wound around the bottom of the inside of a 300 ml rimless Erlenmeyer flask, which is then covered with a metal cap and sterilized by autoclaving at 121° for 15 min. The inoculum is prepared by emulsifying the growth from a slant of an overnight culture (37°) on Brain Heart Infusion Agar in 2 ml of sterile 0·01M phosphate buffer pH 6·8. The flask is inoculated by adding 8·0 ml of the sterile buffer solution and 2·0 ml of the bacterial suspension. Flasks are incubated at 37° for 48 h on a rotary shaker set to revolve at 120 revs/min. The culture is recovered from the flask with a pipette and centrifuged to remove the cells. Sodium merthiolate (0·002%) is added to the clear supernatant which is then stored at −20°C until required for toxin assay.

Extraction of Enterotoxins from Foods

Foods implicated in staphylococcal food poisoning may contain 0·01 μg or less, of enterotoxin/g of food (Casman and Bennett, 1965; Bergdoll, 1972). As the gel-diffusion methods used to detect the enterotoxins have a maximum sensitivity of 0·1–0·5 μg/ml, it is necessary to extract and concentrate the toxin present in a food homogenate in order to detect low concentrations of toxin in a suspect food. Suitable extraction procedures giving the required concentration have been described by Casman and Bennett (1965) and Casman (1967).

The method described by Casman (1967) has proved suitable for the extraction of enterotoxin from both high and low protein foods. In this method, which takes at least 4 d to complete, the enterotoxin is extracted from the food with a sodium chloride solution, lipids are removed with chloroform, and the toxin separated from some of the other proteins present by acid precipitation and column chromatography. The eluate from the column is then concentrated to 0·4 ml.

Method (modified from Casman's (1967) procedure)

1. Extract toxin from suspect food by homogenizing a 100 g sample (if available) in 500 ml of 0·2M NaCl and adjust slurry to pH 7·5. After standing at room temperature for 15 min check pH and adjust if necessary. Centrifuge at 20,000 **g** for 15 min at 5°. Retain supernatant and re-extract solids with a further 125 ml of the salt solution. Pool supernatants.

2. Remove fats by adding $\frac{1}{4}$ vol of chloroform, shake vigourously 10 times through a 90° arc. Centrifuge at 6,000 **g** for 10 min at 5° and separate the aqueous phase. Repeat chloroform extraction if food has a high fat content.

3. Reduce salt content and concentrate extract to about 30 ml by dialysis in Visking Tubing against 30% polyethylene glycol (M.W. 20,000) at 4°. This stage can be carried out overnight. Wash off polyethylene glycol from the outside of the tubing with water, and soak tubing in distilled water for 10 min. Remove the contents from the tubing and wash out tubing with 10 ml of saline. Add washings to concentrated extract and, if necessary, dilute to 40 ml with distilled water.

4. Adjust concentrate to pH 4·5 with 5N HCl, and centrifuge at 20,000 **g** for 15 min at 5°.

5. Remove supernatant and adjust to pH 5·7 with N NaOH solution.

6. Extract supernatant with $\frac{1}{4}$ vol of chloroform and centrifuge as in (2) above. Repeat extraction if a heavy precipitate forms with first chloroform treatment. Re-centrifuge aqueous phase if turbid.

7. Dilute aqueous phase with 40 vols of 0·005 M phosphate buffer pH 5·7; and adjust back to pH 5·7, if necessary, with either 0·005 M H_3PO_4 or 0·005 M Na_2HPO_4. This reduces the molarity of the solution to between 0·005 and 0·01 M.

8. Pass solution through a column (see below) of Watman CM 52 carboxy-methyl cellulose (CMC) ion-exchanger (Reeve Angel and Co. Ltd., 14 New Bridge Street, London E.C.4.) at a flow rate of c. 1·5 ml/min at 4°. This stage can be carried out overnight and the column prevented from drying out by raising the end of the exit tubing from the column above the level of the CMC bed (see Fig. 1). The CM 52 CMC does not require

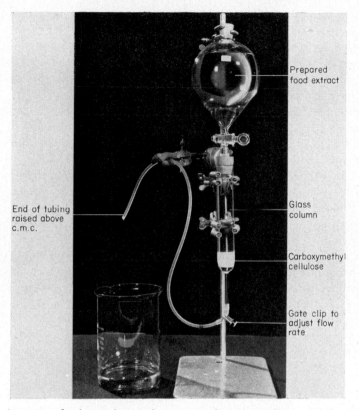

FIG. 1. Apparatus for ion-exchange chromatography using carboxymethyl cellulose to separate enterotoxins from some of the other proteins in a food extract.

the pre-cycling procedure necessary for dried exchanges. The wet slurry is equilibrated against the starting buffer—0·01 M phosphate buffer pH 5·7—and the fines removed following the method recommended by the manufacturers. Adequate columns can be prepared in No. 3 sintered glass mercury filters (J. A. Jobling and Co. Ltd., Sunderland, England) fitted with a piece of silicone tubing and a gate clip to control the flow rate. Completely fill, by suction, the space below the filter with equilibrating buffer. Mount the column vertically and half fill with buffer. Add sufficient of the equilibrated CMC slurry to give a final CMC bed volume of 20 ml (approximate height in a 3·0 cm diam column is 2·5 cm). Pack the CMC by passing 200 ml of equilibrating buffer through it. Do not let the exchanger run dry. Fill column with buffer and insert separating funnel fitted with rubber bung into the top of the column (see Fig. 1). Fill the funnel with prepared food extract, open gate clip and adjust flow rate to 1·5 ml/min.

9. After passing the sample through the CMC wash the exchanger by passing 100 ml of equilibrating buffer through it.

10. Adsorbed proteins, including the enterotoxin, are eluted from the CMC with 150 ml of 0·2 M phosphate buffer pH 7·4 containing 0·2 M NaCl at a flow rate of 2 ml/min. Use positive air pressure to force the last few ml of buffer from the CMC.

11. Concentrate the eluted sample to < 0·4 ml by dialysis in Visking Tubing against 20% PEG at 4°. Wash tubing in water to remove PEG and soak in 0·02 M phosphate buffer pH 7·2 for 15 min before collecting the concentrated sample. 4″ blunt ended forceps with their points and arms covered in rubber tubing are ideal for squeezing the concentrate to one end of the visking tubing, and also for removing PEG from off the outside of the tubing.

12. Centrifuge to remove any insoluble material and extract with chloroform until no further precipitate forms.

Further concentration of the extract may be achieved by lyophilization, but if the sample (neat and serial dilutions) is to be examined for the presence of all enterotoxins by the microslide method (in duplicate) then c. 0·4 ml of material is required.

Enterotoxin recovery from control samples to which known amounts of toxin have been added range from 20 to 80%, depending upon the type of food and the initial toxin concentration.

Production of Antisera

Specific antisera to each of the enterotoxins have been produced by using either highly purified toxin as the antigen, or by absorbing out non-enterotoxinogenic antibodies from an antiserum produced against partly purified material. (Bergdoll, Surgalla and Dack, 1959; Casman, 1960; Casman and Bennett, 1964; Bergdoll, Borja and Avena, 1965; Casman et al., 1967). Different methods for the hyperimmunization of rabbits has been used by these workers. Immunization of rabbits with enterotoxoids gives poor antiserum titres and is not recommended (Bergdoll, 1970).

We obtained high titred antisera to enterotoxins using the following method. Rabbits with low antibody (precipitin) titres to S. aureus are chosen for immunization. Increasing doses of 0·1, 0·5, 1·0, 2·5, 10, 50, 250, 1000 and 2000 μg of purified (> 95%) enterotoxin are used. Each dose of toxin in 1 ml of phosphate buffer pH 7·4 is sterilized using a Millipore millex filter unit, (Millipore (U.K.) Ltd., Heron House, 109 Wembley Hill Road, Wembley, Middlesex, England), and mixed into 1·25 ml of Freund's Complete adjuvant (Difco) to give a stable emulsion. The emulsion can be prepared by pumping the mixture repeatedly through a 2·5 ml plastic disposable syringe (without needle) into a bijou bottle. The emulsion is stabilized when a drop placed on to water remains as a globule and does not spread out over the surface.

The emulsion is then injected subcutaneously into the loose tissue above the rabbits shoulders. Prior to injection, hair is removed from the injection area and the exposed skin swabbed with tincture of iodine. Successive doses are given at about three week intervals depending upon the recovery of the animals body weight and upon repair of the tissue response to the adjuvant. Increase in antibody titre is followed during the latter stages of immunization, and to conserve purified toxin, only rabbits showing the best response are given the final doses. Antibody titres usually reach their peak 7–14 d after injection. Blood may be collected from the marginal ear vein and allowed to clot, when over 100 ml of blood can be collected by successive bleedings over a two week period. After resting the animals, a booster dose of toxin can be given followed by further bleedings.

The separated serum may be preserved with 0·005% merthiolate and stored at −20° in small amounts of 0·2–0·5 ml, but for the long term storage of large volumes of antisera freeze drying is essential.

Serological methods

Precipitin Techniques

Several preciptin methods have been described for the detection and/or quantitation of enterotoxins. Of these methods, the Micro Double-diffusion slide technique and the Tube Double-diffusion technique have been used more frequently than the others for the demonstration of enterotoxins in foods and culture supernatants.

The Micro Double-diffusion Agar Slide Technique (Crowle, 1958)

The slide modification of the Ouchterlony (1958) double-diffusion plate method was described by Wadsworth (1957) and by Crowle (1958). The application of this method to the detection of enterotoxins was first discussed by Casman and Bennett (1964). More detailed information was later given by Hall, Angelotti and Lewis (1965) and by Casman *et al.* (1969).

The method requires only small quantitites (< 0·02 ml) of antiserum and test material, which is very advantageous due to the difficulty and cost of producing specific antiserum.

Specificity

The method gives proven identity of an enterotoxin in the test material by the coalescence of the precipitin line, produced by the interaction with specific antibody, with that of the control toxin in an adjacent well (Crowle, 1961).

Sensitivity

Between 0·1 and 0·2 μg/ml of enterotoxins A and B will just form a stainable precipitin line against the optimal dilution of antibody. However, a concentration of 0·5 μg/ml of toxin is necessary for the precipitin line to coalesce with 1 μg/ml of control toxin in an adjacent well.

Advantages

Only 0·02 ml of reactants are required. Other proteins from foods or culture supernatants do not interfere with the precipitin reaction.

Disadvantages

Low concentrations of toxin < 0·5 μg/ml require 3 d incubation before they can be detected. The preparation of slides is tedious and some practical expertise is necessary to obtain consistent results.

Method

1. Clean polished $3 \times 1''$ microscope slides are coated in agar by immersing them into a near boiling solution of 0·2% (w/v) Difco Bacto Agar. The slides are air dried in a vertical position and stored in dust free conditions.
2. Three pieces of triple layered, 1 cm wide self-adhesive PVC tape (Lassovic Tape, T. J. Smith and Nephew Ltd.) are placed across the slides so that two troughs 17 mm wide and 0·45 mm deep are formed. (see Fig. 2.).
3. Plastic matrices (Casman *et al.*, 1969) are coated on their underside with a film of Repelcote (2% dimethyldichlorosilane in carbon tetrachloride, obtainable from Hopkins and Williams Ltd.).
4. Ten ml of molten 0·5% (w/v) agarose (Miles-Serevac) in 0·145 M NaCl, at 60°, is mixed with 8 ml of 0·04 M Sorensen's phosphate buffer (pH 7·2) in 0·145 M NaCl containing 0·002% merthiolate, at the same temperature. About 0·4 ml of this mixture is pipetted into each trough and covered with

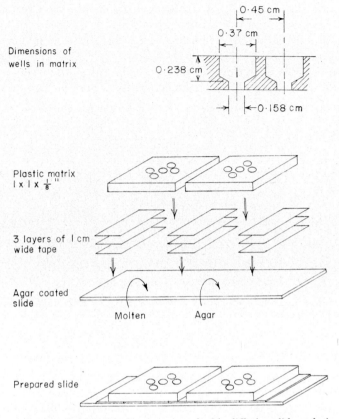

Dimensions of wells in matrix

0·45 cm

0·37 cm

0·238 cm

0·158 cm

Plastic matrix
1 x 1 x ⅛"

3 layers of 1 cm wide tape

Agar coated slide

Molten   Agar

Prepared slide

Fɪɢ. 2. Preparation of slide for the micro double diffusion slide technique.

the matrix so that no air bubbles are trapped in the agar, the excess agar being expelled from the edge of the slide. When set the slides are placed into a moist chamber to avoid drying out.

5. The centre well of the matrix is filled with appropriately diluted antiserum (see below). One outer well or two opposing wells are filled with homologous control toxin at a concentration of 1 μg/ml (see below). The remaining wells are filled with test material. The slides, in the moist chambers, are incubated at 25° for 3 d to allow the precipitin lines to form.

6. Slide off the matrix. Soak the slides, for 20 min each, in two changes of 0·145 ᴍ NaCl and two changes of distilled water. When test materials with high protein contents are examined it is preferable to soak the slides in saline overnight to remove unprecipitated proteins. If these proteins are not removed from the gel they will be stained by the dye, and may obscure weak reactions.

7. Stain the slides in 0·1% (w/v) thiazine red in 1% (v/v) acetic acid for 10 min. Decolourize in two or three changes of 1% (v/v) acetic acid, until the background is almost colourless. Dry the slides at 60° or at room temperature.

8. Examine slides for red precipitin lines (see Fig. 3.) between centre and peripheral wells using oblique lighting against a dark background.

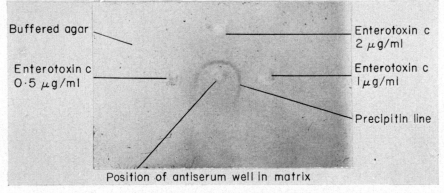

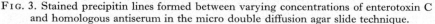

Buffered agar

Enterotoxin c 0·5 μg/ml

Enterotoxin c 2 μg/ml

Enterotoxin c 1 μg/ml

Precipitin line

Position of antiserum well in matrix

FIG. 3. Stained precipitin lines formed between varying concentrations of enterotoxin C and homologous antiserum in the micro double diffusion agar slide technique.

Titration of antiserum

Increasing dilutions of antiserum are reacted against 1 μg/ml of the homologous enterotoxin, both are diluted in 10% Brain Heart Infusion Broth, pH 7·2 (0·37 g % dehydrated powder) containing 0·002% merthiolate. The optimal dilution of antiserum is that which gives a sharp precipitin line mid-way between the inner and outer wells. A measure of the range of toxin concentrations that the optimal dilution of antibody will react with to form an observable precipitin line should be determined with toxin concentrations of between *c.* 5 and 0·1 μg/ml.

Stock solutions of antigen and antibody are kept either as freeze dried material or frozen at —20° in small volumes suitable for dilution to give the working standards. Working standards should also be stored in about 0·5 ml amounts to avoid repeated freezing and thawing which will result in loss of potency. Current working solutions can be stored at —4° for up to 4 weeks without loss of titre.

Double diffusion tube technique (Oakley & Fulthorpe, 1953)

In the system described by Oakley and Fulthorpe (1953), the antibody from the bottom agar layer and antigen from the upper aqueous layer diffuse into the middle agar layer forming a precipitin line in the area of optimum concentrations (see Fig. 4). This method was first used to detect enterotoxins by Bergdoll, Surgalla and Dack (1959).

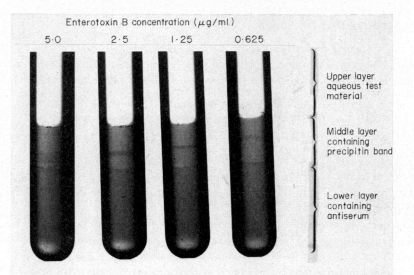

FIG. 4. Oakley and Fulthorpes double diffusion tube technique showing precipitin bands formed between decreasing concentrations of enterotoxin B and homologous antiserum.

Sensitivity

Sensitivity depends upon the thickness of the middle agar layer, relative concentrations of antigen and antibody and incubation time (Oakley and Fulthorpe, 1953). With the procedure described below, 0·6 μg/ml of enterotoxin can be detected after 7 d incubation. It is claimed that 0·05 μg/ml of enterotoxin can be detected after prolonged incubation (Hall, Angelotti and Lewis, 1965).

Specificity

It is not possible to compare directly test samples with enterotoxin controls as in the micro-slide technique. The method relies upon using a specific antiserum.

Advantages

Easy to prepare and handle. Prepared tubes containing only the lower antibody layer can be stored for several weeks at room temperature. We have found that the sensitivity of this method using enterotoxin B is unaffected by sodium chloride concentrations up to 8% (w/v) and by pH of the toxin containing solution between pH 5·0 and 8·5.

Disadvantages

It takes up to 7 d for the precipitin bands to form with low concentrations of enterotoxin. The method is more expensive on antiserum than the micro-slide technique. Some food extracts cause cloudiness of the middle layer which may obscure weak reactions.

Method

1. Agarose, 0·5% w/v, (Miles-Serevac, Maidenhead, Berks, England), in 0·145 M NaCl is dissolved by boiling and filtered whilst hot through a clarifying filter. It is then dispensed in 10 ml amounts and sterilized at 15 lb/sq in (121°) for 15 min.
2. Antiserum is diluted in 0·04 M Sorensen's phosphate buffered (pH 7·2) saline containing 0·002% merthiolate to give the required dilution when mixed with an equal volume of agarose.
3. Molten agarose, at 60°, is mixed with the diluted antiserum and 0·5 ml amounts pipetted into $3 \times \frac{1}{4}''$ tubes as quickly as possible. It is important to avoid leaving traces of antibody agar in the upper parts of the tube. The tubes are sealed with rubber bungs until required.
4. When the tubes are required for use, 0·2 ml of molten, cooled 0·25% (w/v) agarose in phosphate buffered (pH 7·2), saline is layered above the anti-serum base and allowed to set.
5. Known amounts (0·5 ml) of doubling dilutions of the test material in 0·02 M phosphate buffered (pH 7·2) saline are then pipetted over the middle layer and the tubes stoppered. This should be completed within 2 h of adding the middle layer.

Tubes are incubated at 25° for up to 14 d and examined periodically for precipitin line formation.

Titration of antiserum

Increasing dilutions of antiserum in agar (bottom layer) are reacted against a standard concentration of toxin (0·6 μg/ml) diluted in 0·02 M phosphate buffered (pH 7·2) saline. The lowest dilution of antiserum which allows the formation of a sharp precipitin line in the centre of the middle layer after the required incubation time (7 d) is used.

Electroimmunodiffusion

In this technique (Laurell, 1966) the antigen migrates from its application well through agar containing antiserum under the influence of an electric field, a cone shaped precipitin line being formed by the reaction of the toxin with its homologous antibody (see Fig. 5). The method requires the antigen to migrate more rapidly than the antibody and that both migrate towards the same electrode. The height of the precipitin cone is proportional to the concentration of antigen and inversely proportional to the antibody concentration.

Chugg (1972) described electrophoretic conditions under which quantitative results could be obtained at enterotoxin concentrations up to 200 μg/ml. We have modified these conditions so that improved quantitation can be obtained with toxin concentrations between 0·5 and 10 μg/ml. Further modifications of Chugg's (1972) method have recently been described by Gasper, Heimsch and Anderson (1973).

Electrophoresis is carried out in an electrophoresis tank containing a cooling platten (see Fig. 6). The gel is formed in a Perspex mould with cover glass to give a gel 6·5 cm long, 2·5 cm wide and 0·15 cm thick.

Electrophoresis Conditions

Electrophoresis buffer: 0·025 M Veronal buffer (pH 7·4) containing 4 mM calcium lactate. Gel: 1 vol of 2% (w/v) agarose mixed with 1 vol of the buffer solution above containing the required concentration of antibody. This gives a discontinuous buffer system which improves resolution. Temperature: in cooling platten, 20°. Voltage: 150 V (constant). Current: 5·5 to 7·0 mA/slide. Running time: 2 h.

Addition of 1/100 normal rabbit serum to the gel improved resolution and sensitivity.

Sensitivity

Enterotoxins A and B can be detected in concentrations of 0·5 μg/ml.

Specificity

Specific antiserum necessary for proven identity.

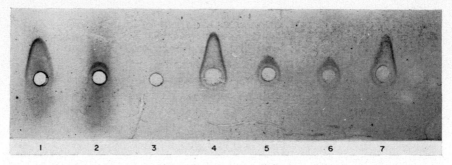

FIG. 5. Precipitin lines formed by electroimmunodiffusion of food extracts containing enterotoxin B and standards in a gel containing enterotoxin B antiserum.
Well 1, Milk extract containing 10 µg/ml of enterotoxin B;
Well 2, Milk extract containing 1 µg/ml of enterotoxin B;
Well 3, Enterotoxin A 10 µg/ml;
Well 4, Meat extract containing 12 µg/ml enterotoxin B;
Well 5, Meat extract containing 1 µg/ml enterotoxin B;
Well 6, Enterotoxin B 1 µg/ml, and
Well 7, Enterotoxin B 10 µg/ml.

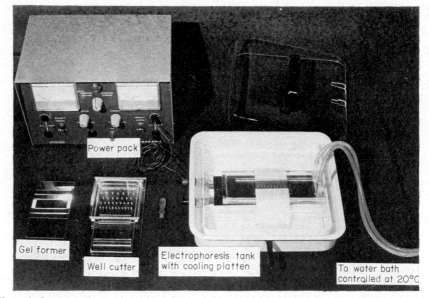

FIG. 6. Immunoelectrophoresis apparatus showing prepared slide on cooling platten in electrophoresis tank.

Advantages

Enterotoxin concentrations above 2 μg/ml can be seen on completion of electrophoresis (2 h) without staining gel.

Disadvantages

Expensive on antiserum. Quantitative results and the demonstration of precipitin lines formed by toxin concentrations between 0·5 and 2μg/ml requires the gels to be stained.

Method

1. Clean, polished 2×3″ Microscope slides are coated in 0·2% (w/v) Difco Bacto Agar and dried in a vertical position.
2. The gel former with cover glass is placed on to an agar coated slide held in a horizontal position. It is filled with the agar/antibody mixture using a Pasteur pipette. The agar/antibody mixture is prepared by mixing 2 ml of molten, cooled (50°) 2% agarose (Miles-Serevac) with an equal volume of 0·025 M Veronal buffer (pH 7·4) containing 4 mM calcium lactate to which has been added the required amount of antibody.
3. When the agar has set, the former is removed and the gel placed on the cooling platten in the closed electrophoresis tank to harden for 15 min. Antigen wells, 2 mm in diam are cut in the gel 0·8 cm apart and 0·5 cm from the anode edge. Wicks connecting the gel to the buffer reservoirs are made of two layers of Watman No. 1 Chromatography paper and are sealed to the edge of the gel with molten, buffered agar.
4. Antigen wells are loaded with 5 μl of control or test material and electrophoresis carried out for 2 h.
5. Gels are washed in 0·2 M NaCl overnight to remove excess antiserum and non-enterotoxigenic proteins from the agar.
6. The agar slide is either dried with a hair-dryer or covered with wet filter paper and dried at 60° in an oven. The slide is stained for 20 min, with a mixture of equal parts of 0·5% Amido Black 10 B in 5% (v/v) acetic acid containing 5% (w/v) $HgCl_2$ and 0·1% Thizaine Red in 1% (v/v) acetic acid. Decolourization of the gel with 1% (v/v) acetic acid is continued until optimal visualization of the precipitin lines is achieved. Rinse slides in distilled water and dry. Measure precipitin cone heights.

Estimation of toxin concentration in unknown samples

Cone height is proportional to toxin concentration. Two standards, 10 and 2 μg/ml, are included on each slide. The log of the cone height of the two standards is plotted against the log of their concentration to give a standard curve. The toxin content of unknown samples is then determined from that dilution of the unknown sample that gives a cone height falling within the two standards.

Haemagglutination methods

Haemagglutination techniques, in which purified antigen or purified antibody is coupled to erythrocytes, are more sensitive methods than the precipitin techniques described previously. Two haemagglutination inhibition techniques (Morse and Mah, 1967; Johnson, Hall and Simon, 1967), in which purified enterotoxin B is attached to sheep erythrocytes, have been described. In the haemagglutination inhibition test system, increasing dilutions of test material are reacted with a constant amount of antibody. Where sufficient enterotoxin is present in the test material the antibody is utilized, so that, upon adding the toxin coated erythrocytes no agglutination occurs, but where free antibody is still available agglutination does take place. Haemagglutination inhibition methods are less sensitive than the Reverse Passive Haemagglutination methods which are the methods of choice (Silverman, Knott, and Howard, 1968). In this method purified antibody is coupled to aldehyde treated erythrocytes with tannic acid and enterotoxin present in the test material causes agglutination of these erythrocytes, by forming antibody—antigen—antibody links from cell to cell.

Reverse passive haemagglutination

Sensitivity
As little as 0·001 μg/ml of enterotoxin can be detected.

Advantages

A rapid test to perform with results in about 4 h once the test system is established. Food extracts can be tested directly without the need for lengthy concentration procedures.

Disadvantages

Expensive on antiserum. A good stock of high titred antiserum is necessary to perform this test routinely. Sensitized cells can be stored for only a week at 4° without deterioration. Difficulty may be experienced in coupling enterotoxin A antisera from individual rabbits to erythrocytes. More experience of this test system with different foods is required.

Method

1. Preparation of IgG immunoglobulin.—The separation of the IgG fraction from whole serum is done using diethylaminoethyl cellulose anion exchanger. Both a batch process (Stanworth, 1960) and a column method (Levy and Sober, 1960) have been successfully used. The separated IgG fraction is concentrated by vacuum dialysis to the same volume as the original serum. Disc electrophoresis of the IgG fraction showed no contamination by other serum proteins.

2. Preparation of sheep erythrocytes (Ling, 1961).—Sheep blood in Alsever's solution (Wellcome Reagents Ltd., Beckenham, Kent, England), is washed three times in normal saline with centrifugation to remove the plasma. 1 vol of 50% suspension of cells in saline is mixed with 1·5 vol of 25% (v/v) pyruvic aldehyde adjusted to pH 7·0 with 1% (w/v) Na_2CO_3 and 0·7 vol of 0·15 M phosphate buffer pH 8·0. The suspension is kept at 4° for 24 h with occasional mixing. Cells are washed six times with saline to remove free aldehyde and stored in saline containing 0·1% (w/v) sodium azide at 4°. Treated cells can be kept for up to six months without deterioration.

3. Sensitization of erythrocytes.—The treated cells are washed twice in saline and a 2·5% suspension of cells in saline prepared. To this is added 1 vol of freshly prepared 1/20,000 tannic acid in saline (see 4) and the mixture is left at room temperature for 15 min. The mixture is then centrifuged at low speed (< 1,000 rev/min) to deposit the cells. Decant the supernatant and wash in 1 vol of saline and centrifuge. To the cells add 1 vol of 0·15 M phosphate buffered (pH 6·4) saline and 1 vol of diluted IgG in the same buffer (see 4 below). Incubate at 37° for 1 h. Centrifuge at low speed and resuspend cells in 1 vol of 0·15 M phosphate buffered (pH 7·2) saline containing 0·8% bovine serum albumen (Fraction V, Armour Pharmaceutical Co. Ltd., Eastbourne, England). Leave at room temperature for 1 h centrifuge slowly and resuspend cells in 0·5 vol of same buffer. Packing the cells too hard during centrifugation can give rise to clumping thereby leading to poor results in the test.

4. Optimum sensitization of cells.—The concentration of tannic acid and IgG immunoglobulin to give optimal sensitization of erythrocytes is determined by treating aliquots of cells with increasing dilutions of tannic acid (1/10,000 to 1/80,000) and treating aliquots of each with increasing dilutions of IgG antibody (1/50 to 1/500) as above. Sensitized cells are then reacted with doubling dilutions (0·1 to 0·00009 μg/ml) of pure toxin as described in the test procedure below. Controls—diluent only—are also included. The optimal concentration of antibody and tannic acid is that showing the highest titre of well agglutinated cells.

Specificity is confirmed by negative reactions against heterologous enterotoxins, and culture supernatants of non-enterotoxinogenic strains of *S. aureus*.

5. Test.—Serial dilutions (1 in 2 or 1 in 10) of test material are prepared in 0·5 ml amounts of phosphate buffered (pH 7·2) saline containing 0·8% bovine serum albumen. To each dilution is added 1 drop (40 drops/ml) of sensitized cells. Mix well and read results after 4 h or overnight. Controls —buffer only—negative control of food extracts etc. are also included (see Fig. 7).

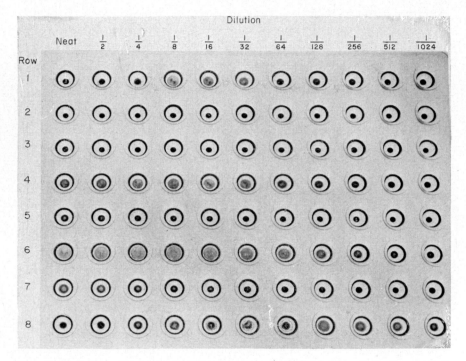

FIG. 7. Detection of enterotoxin B by the reverse passive haemagglutination method using erythrocytes sensitized with enterotoxin B antiserum (IgG fraction).

Row 1, Enterotoxin B, initial concentration 0·1 μg/ml;
Row 2, Enterotoxin A, initial concentration 1·0 μg/ml;
Row 3, Enterotoxin C, initial concentration 1·0 μg/ml;
Row 4, Extract of chicken, containing enterotoxin B;
Row 5, Chicken meat extract, negative control;
Row 6, Extract of prawns, containing enterotoxin B;
Row 7, Prawn extract, negative control, and
Row 8, supernatant Culture of *Staphylococcus aureus* producing enterotoxin B.

Solid phase radioimmunoassay

Two methods have been described recently for the detection of enterotoxin B. (Johnson, Bukovic, Kauffman and Peeler, 1971; Collins, Metzger and Johnson, 1972). In both methods, purified enterotoxin B antibody (IgG fraction) is attached to a solid carrier—polystyrene tubes or bromacetyl cellulose. Quantitative assay is determined by the inhibition of binding ^{125}I labelled antigen to the fixed antibody by the enterotoxin present in the test material. Bound (Johnson et al., 1971) or free (Collins et al., 1972) ^{125}I labelled antigen being subsequently determined using radioactive counting equipment. We have not investigated these methods in this laboratory.

Sensitivity

0·01 to 0·001 μg/ml of enterotoxin B detectable.

Advantages

Rapid (less than 1 d), sensitive, assay method once reagents are prepared.

Disadvantages

Use of radioactive material and expensive counting equipment. Specific antiserum and purified enterotoxin required. Both methods are time consuming in preparation of reagents.

Conclusions

At present it is necessary to concentrate a food extract suspected of containing enterotoxin, so that low levels of toxin can be detected by the precipitin methods currently used. The extraction method described (Casman, 1967) takes 4 d to complete and serological detection a further 1 to 7 d depending upon the method used. More rapid and sensitive methods for detecting enterotoxins have been proposed and of these the reverse passive haemagglutination and solid phase radioimmunoassay methods have the required sensitivity, detecting 0·01 ug/ml or less of enterotoxin. Using these methods, results on foods could be obtained within two working days. However, these methods have only been described for enterotoxins A and B, as they require highly specific antisera and/or purified enterotoxin which are not generally available and very expensive to produce. The reverse passive haemagglutination method has been shown to give non-specific results by some workers (Bennett *et al.*, 1973). It is therefore probable that the radioimmunoassay method or other sensitive and specific technique will be the future methods of choice in suitably equipped laboratories.

Acknowledgements

We wish to thank Prof. M. S. Bergdoll for many discussions over the last 10 years on staphylococcal enterotoxins, and for supplying reagents to initiate our own studies. Thanks also to Prof. A. W. Anderson for advice and Mr A. Liston for his help in developing the electroimmunodiffusion method.

References

BAIRD-PARKER, A. C. (1971). Factors affecting the production of bacterial food poisoning toxins. *J. appl. Bact.*, **34**, 151.

BENNETT, R. W., KEOSEYAN, S. A., TATINI, S. R., THOTA, H. & COLLINS, II, W. S. (1973). Staphylococcal enterotoxin. A comparative study of serological detection methods. *J. Inst. Can. Sci. Technol.*, **6**, 131.

BERGDOLL, M. S. (1967). The Staphyloccocal Enterotoxins, In *Biochemistry of Some Foodborne Microbial Toxins*. (R. I. Mateles and G. N. Wogan, eds). M.I.T. Press, Cambridge, Massachusetts.

BERGDOLL, M. S. (1970). Enterotoxins, In *Microbial Toxins, Vol. III*. (S. J. Ajl, T. C. Montie and S. Kadis, eds). Academic Press, New York.

BERGDOLL, M. S. (1972). The Enterotoxins. In *The Staphylococci*. (J. O. Cohen, ed.). Wiley Interscience, New York.

BERGDOLL, M. S., BORJA, C. R. & AVENA, R. M. (1965). Identification of a new enterotoxin as enterotoxin C. *J. Bact.*, **90**, 1581.

BERGDOLL, M. S., SURGALLA, M. J. & DACK, G. M. (1959). Staphylococcal enterotoxin. Identification of a specific precipitating antibody with enterotoxin-neutralizing property. *J. Immunol.*, **83**, 334.

CASMAN, E. P. (1960). Further serological studies of staphylococcal enterotoxin. *J. Bact.*, **79**, 849.

CASMAN, E. P. (1967). Staphylococcal food poisoning. *Health Lab. Sci.*, **4**, 199.

CASMAN, E. P. & BENNETT, R. W. (1963). Culture medium for the production of staphylococcal enterotoxin. A. *J. Bact.*, **86**, 18.

CASMAN, E. P. & BENNETT, R. W. (1964). Production of antiserum for staphylococcal enterotoxins. *Appl. Microbiol.*, **12**, 363.

CASMAN, E. P. & BENNETT, R. W. (1965). Detection of staphylococcal enterotoxin in food. *Appl. Microbiol.*, **13**, 181.

CASMAN, E. P., BENNETT, R. W., DORSEY, A. E. & ISSA, J. A. (1967). Identification of a fourth staphylococcal enterotoxin, enterotoxin D. *J. Bact.*, **94**, 1875.

CASMAN, E. P., BENNETT, R. W., DORSEY, A. E. & STONE, J. E. (1969). The micro-slide gel double diffusion test for the detection and assay of staphylococcal enterotoxins. *Health Lab. Sci.*, **6** 185.

CASMAN, E. P., BERGDOLL, M. S. & ROBINSON, J. (1963). Designation of staphylococcal enterotoxins. *J. Bact.*, **85**, 617.

CHUGG, L. R. (1972). Rapid sensitive methods for enumeration and enterotoxin assay for *Staphylococcus aureus*. Ph.D. Thesis, Oregon State University, U.S.A. *Dissert. Abs.*, **32**, (8), 4383 B.

COLLINS, W. S., METZGER, J. F. & JOHNSON, A. D. (1972). A rapid solid-phase radioimmunoassay for staphylococcal B enterotoxin. *J. Immunol.*, **108**, 852.

CROWLE, A. J. (1958). A simplified micro double-diffusion agar precipitin technique. *J. Lab. Clin. Med.*, **52**, 784.

CROWLE, A. J. (1961). *Immunodiffusion*. Academic Press, New York.

DIETRICH, G. G., WATSON, R. J. & SILVERMAN, G. J. (1972). Effect of shaking speed on secretion of enterotoxin B by *Staphylococcus aureus*. *Appl. Microbiol.*, **24**, 561.

DONNELLY, C. B., LESLIE, J. E., BLACK, L. A. & LEWIS, K. H. (1967). Serological identification of enterotoxigenic staphylococci in cheese. *Appl. Microbiol.*, **15**, 1382.

GASPER, E., HEMISCH, R. C. & ANDERSON, A. W. (1973). Quantitative detection of type A staphylococcal enterotoxin by Laurel electroimmunodiffusion. *Appl. Microbiol..*, **25**, 421.

HÁJEK, V. & MAŘSALEK, E. (1973). The occurrence of enterotoxigenic *Staphylococcus aureus* strains in hosts of different animal species. *Zbl. Bakt. Hyg.* **223**, 63.

HALL, H. E., ANGELOTTI, R. & LEWIS, K. H. (1965). Detection of staphylococcal enteroxins in food. *Health Lab. Sci.*, **2**, 179.

HOLLANDER, H. O. (1965). Production of large quantities of enterotoxin B and other staphylococcal toxins on solid media. *Act. Pathol. Microbiol. Scand.*, **63**, 299.

JARVIS, A. W. & LAWRENCE, R. C. (1970). Production of high titres of enterotoxins for the routine testing of staphylococci. *Appl. Microbiol.*, **19**, 698.

JOHNSON, H. M., BUKOVIC, J. A., KAUFFMAN, P. E. & PEELER, J. T. (1971). Staphylococcal enterotoxin B: Solid-phase radioimmunoassay. *Appl. Microbiol.*, **22**, 837.

JOHNSON, H. M., HALL, H. E. & SIMON, M. (1967). Enterotoxin B: serological assay in cultures by passive haemagglutination. *Appl. Microbiol.*, **15**, 815.

KATO, E., KHAN, M., KUJOVICH, L. & BERGDOLL, M. S. (1966). Production of enterotoxin A. *Appl. Microbiol.*, **14**, 966.

LACHICA, R. V. F., WEISS, K. F. & DEIBEL, R. H. (1969). Relationships among coagulase, enterotoxin, and heat stable deoxyribonuclease production by *Staphylococcus aureus*. *Appl. Microbiol.*, **18**, 126.

LAURELL, C. B. (1966). Quantitative estimation of proteins by electrophoresis in agarose gel containing antibodies. *Ann. Biochem.*, **15**, 45.

LEVY, H. B. & SOBER, H. A. (1960). A simple chromatographic method for preparation of gamma globulin. *Proc. Soc. Exp. Biol.*, **103**, 250.

LING, N. R. (1961). The attachment of proteins to aldehyde-tanned cells. *Brit. J. Haemat.*, **7**, 299.

MINOR, T. E. & MARTH, E. H. (1972). *Staphylococcus aureus* and staphylococcal food intoxications. A review. II Enterotoxins and epidemiology. *J. Milk Food Technol.*, **35**, 21.

MORSE, S. A. & MAH, R. A. (1967). Microtiter haemagglutination-inhibition assay for staphylococcal enterotoxin B. *Appl. Microbiol.*, **15**, 58.

OAKLEY, C. L. & FULTHORPE, A. J. (1953). Antigenic analysis by diffusion. *J. Pathol. Bacteriol.*, **65**, 49.

OUCHTERLONY, O. (1958). Diffusion-in-gel methods for immunological analysis. *Progress Allergy*, **5**, 1.

REISER, R. F. & WEISS, K. F. (1969). Production of staphylococcal enterotoxins A, B, and C in various media. *Appl. Microbiol.*, **18**, 1041.

SILVERMAN, S. J., KNOTT, A. R. & HOWARD, M. (1968). Rapid, sensitive assay for staphylococcal enterotoxin and a comparison of serological methods. *Appl. Microbiol.*, **16**, 1019.

ŠIMKOVICOVÁ, M. & GILBERT, R. J. (1971). Serological detection of enterotoxin from food poisoning strains of *Staphylococcus aureus*. *J. Med. Microbiol.*, **4**, 19.

STANWORTH, D. R. (1960). A rapid method of preparing pure serum gamma globulin. *Nature, Lond.*, **188**, 156.

UNTERMAN, F. (1972). Comparison of cultivation methods for the detection of enterotoxins produced by staphylococci. *Zbl. Bakt. Hyg.*, **219**, 435.

WADSWORTH, C. (1957). A slide microtechnique for the analysis of immune precipitates in gel. *Internat. Arch. Allergy*, **10**, 355.

Assay of Myo-inositol using the Yeast
Kloeckera apiculata (K. brevis)

R. W. WHITE AND M. E. BLACK

Biochemistry Department, A.R.C. Institute of Animal Physiology, Babraham, Cambridge, England

Myo-inositol occurs free and in combined form in plants, animals and micro-organisms; it was first found in animal muscle over a hundred years ago (Scherer, 1850). It is an essential nutrient for some micro-organisms and notably for human cell lines and is sometimes classed as a vitamin (Wagner and Folkes, 1964). Detailed accounts of its distribution and biochemistry are given in a comprehensive collection of review papers edited by Eisenberg, Albertson and Krauss (1969).

Myo-inositol can be estimated by chemical means, and an enzyme micro-method described by Weissbach (1958) is currently in use (Sundaram, 1972). However, the yeast assay is convenient and highly specific. An assay using yeasts was first suggested by Williams, Stout, Mitchell and McMahon (1941). Burkholder, McVeigh and Moyer (1944) discussed the requirements of the yeast *Kloeckera brevis* for six vitamins and described a method using the growth response of this yeast for assay of inositol. Northam and Norris (1952) studied inositol assays, using *K. brevis* and *Schizosaccharomyces pombe*, with particular reference to statistical analysis. They used the Spekker absorptiometer, but with neutral grey-green filters, to obtain readings of yeast growth density. Campling and Nixon (1954) and Hartree (1957) described the use of the tube assay with *K. brevis* in studies on inositol in foetal fluids and seminal plasma. The method described here is substantially, and the assay medium is precisely as described by Campling and Nixon (1954).

Methods and Materials

Test organism

Yeast *Kloeckera apiculata* NCYC No. 328 (previously *K. brevis*); growth temperature 22–25°.

Maintenance medium

Yeast Extract (Oxoid)	1·5 g
Lab. Lemco	1·5 g
Peptone	2·5 g
Glucose	5·0 g
Distilled Water	500 ml

Stiffen with 1% (w/v) agar, pH is *c.* 5·3 and requires no adjustment. Sterilize in screw-capped bottles and use as slopes. The yeast is subcultured 3 days before assay use, or every one or two months. Slope cultures are stored at room temperature.

Assay medium

ANALAR reagents are used where available, and glass distilled water is used throughout. Three solutions are prepared as follows:

Solution I

Glucose	40 g
Potassium di-hydrogen phosphate (KH_2PO_4)	3 g
DL-Asparagine	4 g
Calcium chloride ($CaCl_26H_2O$)	0·98 g
Magnesium sulphate ($MgSO_47H_2O$)	1 g
Potassium iodide	0·4 ml of a 0·05% (w/v) solution
Ammonium sulphate	4 g
Distilled water	750 ml

This solution must be prepared immediately before use in a litre volumetric flask to allow for subsequent dilution to 1 litre.

Solution II

Boric acid (H_3BO_3)	0·1 g
Zinc sulphate ($ZnSO_4.7H_2O$)	0·04 g
Ammonium molybdate ($(NH_4)_6M_7O_{24}.4H_2O$)	0·02 g
Manganese sulphate ($Mn_2SO_4.4H_2O$)	0·04 g
Copper sulphate ($CuSO_4.5H_2O$)	0·045 g
Ferrous sulphate ($FeSO_4.7H_2O$)	0·25 g
Distilled water to	1000 ml

This solution can be stored refrigerated ($+2°$) for two months.

Solution III

Thiamine	10 mg
Pyridoxine	10 mg

Calcium pantothenate	10 mg
Nicotinic acid	10 mg
Biotin	1 ml of a 10 μg/ml
	solution
Riboflavin	0·5 mg
Distilled water to	100 ml

This solution has been successfully used after 2 weeks refrigerated ($+2°$) storage but it may be conveniently freshly prepared from 1% frozen ($-20°$) stored solutions of the components. Thiamine has been found to be unstable unless stored frozen ($-20°$)—see later.

Whole assay medium

To the 750 ml of Solution I add 2 ml of Solution II and 4 ml of Solution III. Bring to pH 5·0 by the addition of 2·5 ml NNaOH, and make to 1000 ml with distilled water. Mix well, and filter through Whatman No. 1 paper for immediate use.

Inositol standard

An accurate 2 mg/ml solution in glass distilled water is prepared in a volume of 50 ml and stored in the refrigerator. For each test, 1 ml of this solution is diluted to 1000 ml in a volumetric flask, with glass distilled water immediately before use. In the diluted solution the inositol concentration falls on storage, possibly owing to the adsorption onto the glass surface.

If the Solution III of the assay medium is stored refrigerated (not frozen) for periods longer than two weeks, irregular growth of the test organism results. We investigated this by successive replacement of components and found that thiamine was usually the unstable essential substance. *Kloeckera brevis* has a requirement for thiamine and has also been used for its assay, with substantially the same media referred to earlier but thiamine omitted and inositol added (Emery, McLeod and Robinson, 1946; Hoff-Jorgensen and Hansen, 1955). Jones and Finch (1959) described a plate method for assaying thiamine using *K. brevis*.

The Assay

Preparation of samples

One oz screw capped (McCartney) bottles are used. A final volume of 5 ml is made up of 2·5 ml of the whole assay medium and 2·5 ml of standard solution (or test material). Eight levels of standard (including zero) are used; additions are made to the bottles as shown in the table.

TABLE 1. Standards used in the assay of inositol

Bottle no.	Whole assay medium (ml)	Glass distilled water (ml)	Inositol 2 μg/ml (ml)	Total inositol (μg)
1	2·5	2·5	0	0
2	2·5	2·25	0·25	0·5
3	2·5	2·0	0·5	1·0
4	2·5	1·75	0·75	1·5
5	2·5	1·5	1·0	2·0
6	2·5	1·0	1·5	3·0
7	2·5	0·5	2·0	4·0
8	2·5	0	2·5	5·0

Samples are routinely assayed in duplicate. Two additional bottles containing 2·5 ml whole assay medium and 2·5 ml glass distilled water are included, these are not seeded but used for a blank solution in the optical density readings.

Samples must be diluted so that 2·5 ml contain not more than 5 μg inositol. Those suspected of containing very large concentrations of inositol may be assayed through a series of tenfold dilutions. Solutions of unknown material for assay must:

(a) Be water clear
(b) Not precipitate on autoclaving
(c) Be at pH 5·0 and remain so on autoclaving
(d) Contain nothing inhibitory to the test organism, such as heavy metal ions, and traces of organic solvents which are not necessarily totally voided on heating.

All bottles (standard curve and samples) are sterilized by autoclaving at 112°/20 min, observed after cooling for any turbidity and seeded.

Seeding and growth

Seed is prepared by suspending in 10–20 ml of sterile saline, one or two loopfuls taken from a three-day-old agar slope culture of *K. apiculata*, to give a turbidity just visible to the naked eye. One drop of this suspension is added to each assay bottle using a sterile Pasteur pipette and observing aseptic precautions. The seeded test bottles are well mixed and incubated at 25°/3 days, shaking daily.

Reading the test

The turbidity is estimated in terms of optical density on a Spekker H760 absorptiometer, each bottle being read against unseeded control medium in 24 × 5 mm cells (5 mm light path) using dark red Kodak No. 8 filters. The readings obtained from each pair of duplicates in the standard curve set are averaged and plotted against total μg of inositol, to give the standard curve. The pair of readings obtained from each unknown fraction are averaged, and the amount of inositol present read off the standard curve. Where duplicates show a divergence greater than 0·05 on the Spekker absorption scale, sample assays are repeated if possible.

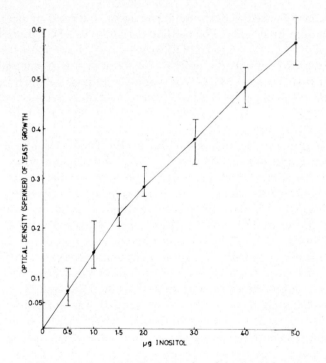

FIG. 1. Means of 10 standard curves taken at random from inositol assays over a period of 5 years.

The yeast has an absolute requirement for inositol in the presence of other essential substances. The assay method is very reproducible. Figure 1 shows the means and limit of a series of 10 standard curves taken at random over a five-year period.

Cleaning glassware

Separate glassware is kept for inositol assays. Neither soap nor ordinary detergents are ever used. A stock cleaning solution is made up as follows:

Sodium phosphate tribasic $Na_3PO_412H_2O$ 500 g
Sodium hexameta-phosphate $(Na_3PO_3)_6$ 120 g
Distilled water   1300 ml

For use, dilute 20 ml to a litre of water. Boil all glassware in this, rinse twice in distilled and glass distilled water.

Preparation of Biological Fluids

Perchloric acid is used as de-proteinising agent, followed by neutralization with potassium hydroxide. The method is as follows: To 1 ml of biological material add 10 ml of 0·6 N-$HClO_4$, usually at 0°, leave for a few minutes and centrifuge to separate the protein. To 10 ml of supernatant add 1 ml of 5 N-KOH then bring to pH 5·0 by gradually adding N-KOH, cool on ice for 30 min, centrifuge for 2–3 min and immediately remove supernatant for assay. Liver, thyroid and other tissue samples have been regularly prepared by this means.

Alternative deproteinizing agents have been tried. Trichloroacetic acid at final concentrations of 1, 3 and 5% (w/v) followed by ether extraction resulted in solutions which precipitated on heating. Treatment with zinc sulphate neutralized with sodium hydroxide as described by Dawson, Elliott, Elliott and Jones (1969) also resulted in solutions which reprecipitated, and caused obvious variations in yeast growth. Samples boiled in acetate buffer (0·1, M pH 4·8) and filtered, gave solutions which sometimes precipitated slightly on autoclaving; also the acetate buffer was found to affect to a small degree the amount of yeast growth obtained in the presence of known levels of inositol. This method could be successfully used provided that the effect on yeast growth was compensated for by substituting the acetate buffer for distilled water in the standard curve bottles, and that no reprecipitation occurred on heating.

Bound inositol for instance as inositides in samples produced by lipid fractionation, is released by hydrolysis with hydrochloric acid. This is carried out for 18 h/105° in 5N HCl in a sealed tube, the acid removed *in vacuo*, the residue made up to a known volume with water and adjusted to pH 5·0 for assay.

Some of the work arising from our experience with the assay is described by Freinkel, Dawson, Ingbar, and White (1959) and Setchell, Dawson and White (1968). Dawson, Freinkel and White (1962) investigated possible sites of inositol uptake in the assay organism. The inositol present in the whole yeasts was only released after mechanical disruption of the yeast cells. The major inositol containing component, a water soluble neutral derivative, was later identified as galactinol (myo-inositol-1-α-D-galactopyranosil) by Dworsky and Hoffmann-Ostenhof (1967).

Prottey, Seidman and Ballou (1970), after culturing *K. apiculata* in the presence of tritiated myo-inositol, found that the major radioactive lipid was ^{3}H-phosphatidyl myo-inositol.

In the absence of inositol, growth of *K. apiculata* ceases, presumably because of the inability of the cells to synthesise the inositol containing components essential for budding and development of daughter cells. It is likely, remembering their tight attachment to the cell, that these components are essential structural, rather than metabolic, elements.

References

BURKHOLDER, P. R., McVEIGH, I. & MOYER, D. (1944). Studies on some growth factors of yeasts. *J. Bact.*, **48**, 385.

CAMPLING, J. D. & NIXON, D. A. (1954). The inositol content of foetal blood and foetal fluids. *J. Physiol.*, **126**, 71.

DAWSON, R. M. C., ELLIOTT, D. C., ELLIOTT, W. H. & JONES, K. M. (1969). *Data for biochemical research*, (2nd Ed.), Clarendon Press, Oxford.

DAWSON, R. M. C., FREINKEL, N. & WHITE, R. W. (1962). The fate of *meso*-inositol during the growth of an inositol-dependent yeast Kloeckera brevis. *J. gen. Microbiol.*, **27**, 331.

DWORSKY, V. P. & HOFFMANN-OSTENHOF, O. (1967). Galaktinol (myo-inositol-1-α-D-galaktopyranosil) in Hefearten. *Hoppe-Seyler's Z. Physiol. Chem.*, **348**, 815.

EISENBERG, F., ALBERTSON, P. D. & KRAUSS, M. (1969). Cyclitols and phosphoinositides: chemistry, metabolism and function. *Ann. N. Y. Acad. Sci.*, **165**, Art. 2.

EMERY, W. B., McLEOD, N. & ROBINSON, F. A. (1946). Comparative microbiological assays of members of the Vitamin B complex in yeast and liver extracts. *Biochem. J.*, **40**, 426.

FREINKEL, N., DAWSON, R. M. C., INGBAR, S. & WHITE, R. W. (1959). The free myo-inositol of thyroid tissue. *Proc. Soc. Exp. Biol. Med.*, **100**, 549.

HARTREE, E. F. (1957). Inositol in seminal plasma. *Biochem. J.*, **66**, 131.

HOFF-JORGENSEN, E. & HANSEN, B. (1955). A microbiological assay of Vitamin B. *Acta Chemica Scand.*, 9, 562.

JONES, A. & FINCH, M. (1959). Plate assay of thiamine. *Appl. Microbiol.*, **7**, 309.

NORTHAM, B. E. & NORRIS, F. W. (1952). A microbiological assay of inositol its development and statistical analysis. *J. gen. Microbiol.*, **7**, 245.

PROTTEY, C., SEIDMAN, M. M. & BALLOU, C. E. (1970). Growth of *K. brevis* in the presence of tritiated myo-inositol. *Lipids*, **5**, 463.

SCHERER, D. (1850). Uber eine aus dem muskelfleische gewonnene zuckerart *Annalen der Chemie*, **73**, 322.

SETCHELL, B. P., DAWSON, R. M. C. & WHITE, R. W. (1968). The high concentration of the free myo-inositol in rete testis fluid. *J. Reprod. Fert.*, **17**, 219.

SUNDARAM, T. K. (1972). Myo-inositol catabolism in *Salmonella typhimurium* Enzyme regression dependent on growth history of organism. *J. gen. Microbiol.*, **73**, 209.

WAGNER, A. F. & FOLKES, K. (1964). *Vitamins and co-enzymes* (p. 272), Interscience Publishers, New York.

WEISSBACH, A. (1958). Enzymic determination of myo-inositol. *Biochim. Biophys. Acta*, **27**, 608.

WILLIAMS, R. J., STOUT, A. K., MITCHELL, K. H. & McMAHON, J. R.(1941). Assay method for inositol. *Univ. Texas Pub. No. 4137, 27*.

Notes on the Preservation and Checking of Vitamin Assay Bacteria*

L. B. PERRY, I. J. BOUSFIELD AND J. M. SHEWAN

National Collection of Industrial Bacteria, Torry Research Station, Ministry of Agriculture, Fisheries and Food, Aberdeen, Scotland

The National Collection of Industrial Bacteria (NCIB) at present comprises over 3000 strains of bacteria and bacteriophages, most of which are of general scientific interest for research and teaching purposes or are of known or potential industrial interest. However, amongst the most important cultures in the Collection are those concerned with a variety of microbiological assays. This chapter discusses aspects of the preservation and checking of such cultures with particular reference to the lactic acid bacteria used for vitamin and amino acid assay. The checking of *Lactobacillus casei* var. *rhamnosus* used for folate assay is given as an example. The appendix to this chapter is intended to serve as a reference source of microbiological assays for vitamins.

Preservation of Assay Cultures

Maintenance in active culture

Maintenance in the active state is still a widely used way of preserving assay cultures and no specialized equipment or facilities are necessary. Many assay cultures can be maintained this way easily; for instance as far as we know antibiotic assay cultures have presented few problems due to culture maintenance procedures (Kavanagh, 1963a). However, it has been noted that, for example, *Bacillus subtilis* should be maintained under conditions that do not permit reversion from rough to smooth type or vice-versa, depending on which variant is found more suitable (Lees and Tootill, 1955). F. N. Mulholland (Boots Pure Drug Co., pers. comm.) kindly tested our 3 descendants of Waksman's strain 0 of *Bacillus cereus* (= ATCC 11778) in oxytetracycline assay; of these NCIB 8012 was most sensitive but NCIB 8849 and NCIB 9231 were almost as sensitive. It is probably important to maintain the antibiotic resistant strains in the presence of the antibiotic in question.

In the case of the lactic acid bacteria used for vitamin assay the repeated subculturing which is necessary together with the accumulation of acid can cause difficulties. There are a few detailed reports of variation (e.g. Nymon and Gortner, 1946; Skeggs, 1961; Coultas, Albrecht and Hutchison, 1966; Albrecht and Hutchison, 1969) and some brief ones (e.g. Kodicek and Pepper, 1948; Loy and Wright, 1959; Pearson, 1967) but sometimes the evidence is not complete. Some active cultures have been successfully maintained for many years (e.g. Price, 1967). The amino acid requirements are considered fairly stable (Shockman, 1963), although variation in 'Leuconostoc mesenteroides' P.60 has been described (Shekleton and Haynes, 1959; Hartley, Ward and Carpenter, 1965).

Snell (1950) reported that the experience of various workers indicated that prolonged exposure to high acidity weakened the organisms. One way of preventing high acidity is to incubate the stock cultures only long enough to get distinctly visible growth along the line of the stab and then to transfer them to the refrigerator. Excessive acidity may also be counteracted by buffering the maintenance medium. In addition to the commonly employed buffers 0.3% (w/v) $CaCO_3$ has been used (Nymon and Gortner, 1946) with a low (0.1% w/v) sugar concentration (Lawrence, Herrington and Maynard, 1946; Skeggs, 1963a). Maintenance media include Lactobacilli agar AOAC (Anon, 1970), liver tryptone agar (Nymon and Gortner, 1946), skim milk with 1% tryptose (Skeggs, 1963b, 1967) and many others (e.g. Barton-Wright, 1962; Shockman, 1963). Further information on procedures can be obtained from other sources (e.g. Snell, 1950; Kavanagh, 1963b, Shockman, 1963; György and Pearson, 1967; Price, 1967; Bolinder, 1972) and individual methods.

To grow lactobacilli for freeze-drying the NCIB now usually uses commercially available Lactobacilli MRS Broth (de Man, Rogosa and Sharpe, 1960) because it gives good growth. The serial subculture which might cause difficulties is largely avoided by laying down a freeze-dried, seed stock at $-20°$ for preparation of subsequent freeze-dried batches.

Details of inoculum preparation are given in individual methods but it may be noted that in the case of Lactobacillus leichmannii Skeggs (1963b) recommended daily subculture for several days in assay medium supplemented with 0.1 ng vitamin B_{12}/ml. This gave a more rapid and sensitive response than a milk culture inoculum. However, the organism was not to be thus maintained for longer than 2 weeks because it was said that continued subculture could diminish the capacity of the organism to utilize bound vitamin B_{12} in serum. Other authorities (e.g. Anon, 1970) also recommend several transfers at "short" intervals before using various strains for assay. A technique for using "log phase" inocula has been described (see below); also the use of depleted cells for improved zone definition in plate assays has been reported (Gare, 1968). However, there is still doubt about when and where variation might occur and what media are best for active maintenance. In any event, variation due to long term subculture need not be a problem now because regular recourse to freeze-dried cultures is easy.

Freeze drying

All assay cultures in the NCIB are preserved by freeze-drying. The basic routine used by the NCIB has been described elsewhere (Bousfield and MacKenzie, 1972), but briefly, the main points are as follows. Cultures are grown under suitable conditions, the cells are spun down, re-suspended in *"mist desiccans"* (Fry and Greaves, 1951), dispensed (0·1 ml amounts) into ampoules containing an identification slip and centrifugally freeze-dried. After 2·5 h primary drying, the ampoules are constricted and left overnight on the secondary dryer, after which they are sealed off under vacuum.

For a large culture collection freeze-drying is the most convenient way to preserve assay cultures, but it is being used increasingly in many other laboratories. Freeze-dried cultures take up little space and do not require special storage facilities. They have a long shelf life, so many identical cultures can be prepared at one time. Thus stocks need be renewed only occasionally and maintenance requires only a small effort in contrast to the subculturing routine needed for maintenance in active culture. The undesirable effects of regular and repeated subculturing are also avoided. From a culture collection point of view, freeze-dried cultures are also conveniently transported by mail.

The main disadvantages of freeze-drying are that special equipment and expertise are needed, there may be a high initial 'kill', especially with Gram negative organisms, and if the culture is found to be contaminated after freeze-drying, the whole process must be repeated. In addition to the routine maintenance of stock cultures, freeze-drying has been used with apparent success to store inocula for direct use in various assays (e.g. Scholes, 1961; Gorin, Meulenhoff and Yarrow, 1970).

Freezing

Preservation of stock cultures by freezing is steadily becoming more popular, especially for organisms which freeze-dry poorly or are difficult to maintain in active culture. Dense suspensions of cells in a suitable medium containing a protective agent such as glycerol (Hollander and Nell, 1954; Howard, 1956) are dispensed in to ampoules and are rapidly frozen, usually by immersion in liquid nitrogen or by exposure to very cold nitrogen gas. Alternatively, cultures may be frozen as pellets (Cox, 1968). The frozen cultures are then transferred either to a deep-freeze cabinet or to a liquid nitrogen refrigerator. If cultures are frozen by immersion in liquid nitrogen, it is a wise precaution to use suitable plastic containers, since glass ampoules may crack, especially at the point of sealing, and allow liquid nitrogen to seep into them during immersion. When the ampoule is removed subsequently from the liquid nitrogen, the rapidly expanding nitrogen gas may cause an explosion (Greenham, 1967). However, if cold nitrogen gas is used for freezing (and storage), this problem is not encountered. We use cold storage for certain cultures in the NCIB and both freezing and storage are carried out in the extremely cold gas produced from evaporating liquid nitrogen.

There is a relatively high survival rate for cultures frozen with liquid nitrogen and it is presumed that such cultures will last for a very long time at low temperatures. The main drawbacks are that the low temperature must be maintained continuously, which means continued access to cold storage facilities or a continually available supply of liquid nitrogen. The latter can prove expensive. The transport of cultures may result in thawing which, depending upon the circumstances and the particular organism, may be undesirable or even deleterious.

The use of liquid nitrogen for the preservation of ready-to-use assay inocula was described by Sokolski, Stapert and Ferrer (1964). Earlier reports described the successful preservation of inocula by freezing at higher temperatures (Tanguay, 1959). Storage of packed cells at $-20°$ in a refrigerator has been successful (Bolinder, 1972). Stab stock cultures have been stored at $-20°$ for more than 6 months (Hansen, 1964).

Other methods

There are numerous other methods of preservation described in the literature, almost all of which rely on desiccating the culture in some way. The excellent review by Fry (1954) still serves as a useful introduction to those interested in such methods. Preservation of cultures by methods other than freezing or freeze-drying is probably not widely used for assay organisms, but the gelatin disc method of Stamp (1947) has been successfully used by some workers (e.g. Chanarin, Anderson and Mollin, 1958; Anderson, Peart and Fulford-Jones, 1970).

Checking for Culture Variation

It is desirable that cultures supplied from a culture collection should be authentic. In the microbiological assay field, the NCIB tests those cultures which might show variation, the more widely used of these being given priority as it is impractical to check most of the assay functions. The checking has been confined mainly to folate and vitamin B_{12} assay cultures, to mutants which can be unstable, and to cultures which are the subject of queries by customers. For the cultures we do not check we anticipate that information from users will reveal the presumably rare faults which could occur.

Mutants produced by strong chemical mutagens or by UV radiation are often relatively unstable and require special care. Reversion of auxotrophic mutants is monitored easily by plate methods if the nutritional requirements are not too complex (e.g. Meynell and Meynell, 1965) using single omission tests, incorporating the growth factors in the medium and plating out the culture. Precautions for such mutants include adequate maintenance concentrations of the growth factors and re-isolation from single colonies as necessary.

In the NCIB most attention is paid to the wild-type cultures commonly used for growth factor assays. In such assays there can be many sources of error (e.g. Girdwood, 1963) some of which may not be quickly identifiable (e.g. Skeggs, 1963b). The task of tracing such errors is facilitated if the culture is known to be satisfactory, especially for those workers for whom microbiological assay is only a small part of their job. Quite often we receive queries from such workers and it is part of the NCIB's function to ascertain that, with respect to standard compounds, the cultures in question are satisfactory and to provide information and advice where possible. In our experience the culture is rarely a major problem unless mishandled but on some occasions we have received reports of differences in sensitivity of different subcultures of the same original strain; in one instance this was apparently a major change (see below).

However, if a growth factor assay is not giving a satisfactory standard curve such as those shown in Fig. 1 we feel that the following simple procedure by the user will at least check the assay culture with some confidence. *Lactobacillus casei* var. *rhamnosus* (ATCC 7469 = NCIB 8010 = NCIB 6375 ('*L. casei*')) which is widely used to assay serum folate is given as an example. Eigen and Shockman (1963) noted occasional variation in its sensitivity to pteroylglutamic acid (PGA). The rationale of the way to use growth factor assay cultures in broth has been discussed by Shockman (1963) and others.

The assay culture, freeze-dried or otherwise, should always be checked for purity before use. A method of checking devised by one of us (LBP) and used routinely in the NCIB is described elsewhere (Bousfield and MacKenzie, 1972). After checking for purity, a simple test of the response of the culture to the compound being assayed is desirable. A relatively easy check for '*L. casei*' might be based on the *preliminary* results reported next.

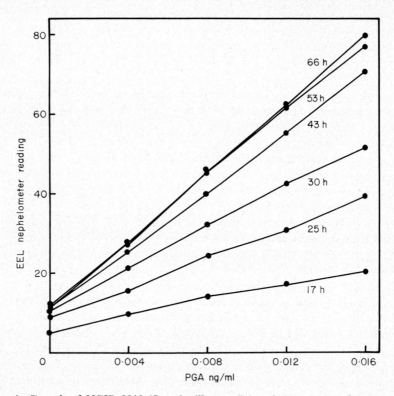

FIG. 1. Growth of NCIB 8010 '*Lactobacillus casei*' in relation to pteroylglutamic acid concentration at various ages (Difco Folic Acid Casei Medium; 37·0°; same tubes measured at intervals).

Replicate optically-matched tubes of assay medium containing a known concentration of PGA (e.g. 0·016 ng/ml) are each inoculated with a precise volume (we use 1 carefully dispensed drop) of a few hours old, weakly turbid log phase culture growing in second or more subculture in assay medium containing 0·12 or 0·16 ng/ml PGA. Controls containing no PGA are also inoculated. The log phase inoculum technique has been described previously (Toennies, Frank and Gallant, 1952; Eigen and

Shockman, 1963; Shockman, 1963). The cultures are incubated in a water bath at 37·0 ± 0·05° and turbidity measurements are made at intervals until growth is complete; a growth curve is then plotted (Fig. 2) and compared with curves derived previously from known satisfactory subcultures. If the colour of the medium affects the turbidity measurements, the cells

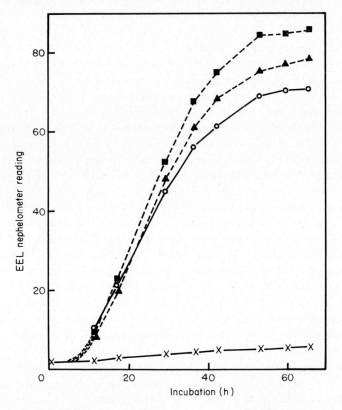

F I G. 2. Growth of NCIB 8010 'Lactobacillus casei' with various folate sources (BBL Folic Acid PGA Broth; closed triangle, 37·0°; same tubes measured at intervals). Closed square, human serum × 1/200 Wellcomtrol Two horse serum × 1/235; open circle, pteroylglutamic acid (PGA) at 0·016 ng/ml, and cross, control without PGA.

can be centrifuged after the final measurement, resuspended in unheated medium and remeasured for better comparison with previous results. The control tube series serves as a check on the absence of folates from the basal medium. New batches of medium can be checked in the same way, using the previous batch as a control. Eigen and Shockman (1963) used the yield of a 0·079 ng/ml PGA medium to determine if an 'L. casei' culture should be replaced. Growth curves for other PGA concentrations are recorded in the studies of Ohara and Silber (1969) and Streeter and O'Neill (1969). Some curves for *L. leichmannii* (ATCC 7830 = NCIB 8118) with vitamin B_{12} are given by Oberzill and Geisler (1970).

We have not tested the effects on the curve of variables such as the number or age of subcultures in log phase inoculum medium, and heating or storage of the checking medium which may cause colour and other changes. However, if conditions are standardized presumably such effects should not matter. Also oxygen can affect the growth rate under some conditions (Manderson and Doelle, 1972) and thus the shaking to resuspend the cells might be important. A separate concentrated stock solution of PGA in NaOH, solely for checking purposes, can be kept for at least 3 months at 0° in the dark (Eigen and Shockman, 1963, 320 μg/ml 0·01 N NaOH; Spray, 1964, 100 μg/ml 0·001 N NaOH).

To indicate that the responses to human serum and to PGA are similar, a suitable dilution of serum can be selected to give a curve close to the PGA test curve (Fig. 2). Although similarity is not proved, it is not contra-indicated. We tested horse serum, now available in freeze-dried form, and obtained a similar curve. Fortunately its strong yellow colour was markedly reduced by the heat extraction procedure of Waters and Mollin (1963), which was used before testing both sera. The extracts were finally membrane-filtered after centrifugation. All tubes contained equal amounts of autoclaved ascorbic acid—phosphate buffer. In the same experiment, each folate source was tested at a second concentration 10% higher or lower than shown in Fig. 2—for clarity the resultant curves have been omitted; in each case the slope and the final reading were greater for the higher concentration. Streeter and O'Neill (1969), and, to a lesser extent, Smillie (1967) found that the rates of the response of '*L. casei*' to PGA and to serum were different, but the conditions used by these workers were not comparable with ours. The study of Fleming, Comley and Stenhouse (1971) included comparisons of assay results for heat extracted and aseptically added sera at 1/100 and 1/200 dilutions.

We have not tested the physiological isomer of 5-methyltetrahydropteroylglutamic acid. Optical density/concentration curves for pteroylpolyglutamates have recently been recorded by Tamura, Shin, Williams and Stokstad (1972). The response to the triglutamate form was always represented by a concave curve in the study of Toennies *et al.* (1952). The specific growth activities of several folates for '*L. casei*' have been noted (Bird and McGlohon, 1972).

It may be of interest that '*L. casei*' can accumulate PGA to some (undetermined) extent and can grow on after the medium is depleted (Toennies, Winegard and Gallant, 1950). Uptake of (^{3}H)folic acid was studied by Cooper (1970). Presumably using the same organism Gare (1968) showed that washing the inocula removed only slowly and incompletely the apparently stored folate. The above log phase technique, requiring no washing and giving no appreciable blank growth at 48 h, does not appear to have been widely tested if at all in serum folate assays although several procedures may approximate to it (Grossowicz, Mandelbaum-Shavit, Davidoff and Aronovitch, 1962; Rae and Robb, 1970; Bird and McGlohon 1972; Tennant and Withey, 1972).

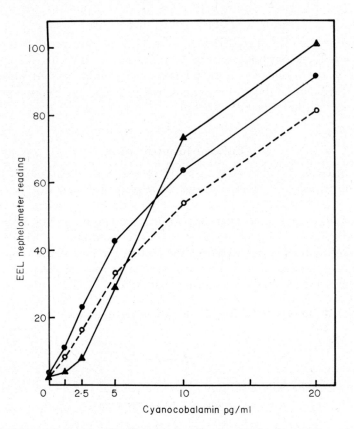

FIG. 3. NCIB 7854 *Lactobacillus leichmannii*, a presumed subculture variant, compared with sibling cultures; growth obtained in relation to cyanocobalamin concentration (Difco B$_{12}$ Assay Medium USP; 37°; 66 h). Closed triangle, NCIB 7854; closed circle, NCIB 8117, and open circles ATCC 4797. The results for the last mentioned came from a different experiment and therefore are not quantitatively comparable.

NCIB 6375 tended to clump more than NCIB 8010 under some of our conditions. Under other conditions (Spray, 1964) the amount of clumping of the two strains varied in different experiments but the responses to

PGA were virtually the same (Dr. G. H. Spray, pers comm). Both cultures are derivatives of ATCC 7469 and the present possible difference presumably arises from their different histories of subculture.

The major apparent change in response referred to earlier is that of NCIB 7854 *L. leichmannii* to cyanocobalamin shown in Fig. 3, its response was sigmoid in contrast to that of NCIB 8117. The 2 cultures are supposedly descendants of ATCC 4797; NCIB 7854 coming to the NCIB via intermediate sources and NCIB 8117 being acceded directly from the American Type Culture Collection (ATCC). The difference was first observed by Dr. I. J. Temperley and Dr. G. H. Spray and they brought it to our attention: our findings (Fig. 3) were similar. It seems likely that the change was a result of culture maintenance procedures but presumably accidental culture substitution cannot be ruled out. Dr. Spray now kindly checks our seed stocks of NCIB 8117.

Culture Identification

When ordering a culture for assay purposes it is important to quote a catalogue number in addition to the name since the latter may be changed or may not be accepted universally. Sometimes, as in the case of *L. leichmannii*, there is more than one strain used commonly for assaying the same compound. Bolinder (1972) stressed the importance of the use of the catalogue number in the literature and cited examples of the re-classification and re-naming of several of the lactic acid assay bacteria. We might add that at least six sources indicate that ATCC 7469 *L. casei* var. *rhamnosus* has at times been known as *L. casei* ε and *L. helveticus* (without catalogue number), especially for the assay of riboflavin.

Appendix

Some microbiological assays for vitamins, and culture collections supplying the microorganisms

The collections of microorganisms maintained in the various countries have been geographically indexed by Martin *et al.* (1972). The UK Collections supplying the vitamin assay cultures in Table 1 are given below, together with the American Type Culture Collection, and others whose catalogues we have.

NCIB   National Collection of Industrial Bacteria, Torry Research Station, P.O. Box No. 31, Aberdeen, AB9 8DG, Scotland.

CCAP   Culture Collection of Algae and Protozoa, 36 Storey's Way, Cambridge CB3 0DT.

TABLE 1. Some microbiological assays for vitamins (Compiled mainly from Kavanagh (1963b, 1972) and György and Pearson (1967)*

Vitamin	Test Organism	NCIB or other UK Collection No.*	ATCC No.*
Biotin	Lactobacillus casei var. rhamnosus	8010	7469
	Lactobacillus plantarum	6376	8014
	Neurospora crassa	CMI 53240	9279
	Saccharomyces cerevisiae	NCYC 79	7754
	Bacillus coagulans	8870	12245
	Lactobacillus casei var. rhamnosus†	10463†	
Folates	Lactobacillus casei var. rhamnosus	8010	7469
	Pediococcus cerevisiae	7837	8081
	Streptococcus faecium	6459, 8123	8043
	Tetrahymena pyriformis	CCAP L 1630/1 W	10542
Inositol	Saccharomyces uvarum	NCYC 74	9080
	Schizosaccharomyces pombe	NCYC 535	16491
Nicotinic acid	Leuconostoc mesenteroides	6992	9135
Nicotinic acid+nicotinamide+nicotinuric acid etc.	Lactobacillus casei var. rhamnosus	8010	7469
	Lactobacillus plantarum	6376	8014
Pantethine	Lactobacillus helveticus	8733	12046
Panthenol	Pediococcus acidilactici	6990	8042
Pantothenic acid	Lactobacillus plantarum	6376	8014
	Saccharomyces uvarum	NCYC 74	9080
Pteridines, unconjugated	Protozoans, various		Yes
Pyridoxal	Lactobacillus casei var. rhamnosus	8010	7469
Pyridoxal+pyridoxamine	Streptococcus faecium φ 51	NCDO 1229	
	Streptococcus faecium	6459, 8123	8043
Pyridoxal+pyridoxamine+pyridoxine	Saccharomyces uvarum	NCYC 74	9080
	Kloeckera brevis		

Vitamin	Test Organism	NCIB or other UK Collection No.†‡	ATCC No.†
Riboflavin	Lactobacillus casei var. rhamnosus	8010	7469
	Streptococcus faecalis var. liquefaciens	7432	10100
	Kloeckera apiculata	NCYC 245	18212
	Kloeckera apiculata		9774
Thiamin	Lactobacillus fermentum	6991	9338
	Lactobacillus viridescens	8965	12706
	Phycomyces blakesleeanus		Yes
	Escherichia coli	8134 = 9270	10799 = 14169
	Euglena gracilis	CCAP 1224/5	12716
Vitamin B_{12}	Lactobacillus lactis	7278	8000
	Lactobacillus leichmannii	8117	4797
	Lactobacillus leichmannii	8118	7830
	Ochromonas malhamensis	CCAP L 933/1a	11532

* A list of additional references to these assays is available from NCIB.

† Culture collection names and addresses are given in this Appendix.

‡ Different cultures of the same strain from different depositors are held under different NCIB numbers (Details on request).

† Chloramphenicol resistant derivative of NCIB 8010.

CMI   Culture Collection, Commonwealth Mycological Institute, Kew, Surrey.

NCDO National Collection of Dairy Organisms, National Institute for Research in Dairying, Shinfield, Reading, Berkshire.

NCYC National Collection of Yeast Cultures, Brewing Industry Research Foundation, Nutfield, Surrey.

ATCC American Type Culture Collection, 12301 Parklawn Drive, Rockville, Maryland 20852, U.S.A.

Centraalbureau voor Schimmelcultures

Fungi: Oosterstraat 1, Baarn, Netherlands.

Yeasts: Yeast Division, Julianalaan 67A, Delft, Netherlands.

Collection de l'Institut Pasteur, 25 rue du Docteur Roux, Paris 15 ème, France. (Bacteria).

Czechoslovak Collection of Microorganisms, J. E. Purkyně University, tř. Obránců míru 10, Brno.

Instytut Przemysłu Fermentacyjnego, Zakład Mikrobiologii Techniczneji Biochemii, Warszawa, ul. Rakowiecka 36.

Institut for Fermentation, 4–54 Juso-nishinocho, Higashiyodogawaku, Osaka, Japan.

Colección de Cultivos Microbianos, Facultad de Farmacia y Bioquímica, Junín 954, 4° piso, Buenos Aires, Argentina.

Addendum

The effect of added buffer salts on serum folate values, differing among closely related *L. casei* var. *rhamnosus* cultures, has recently been reported by Tennant, Newberry, Davies and Dziedzic (*J. Appl. Bact.*, in press).

References

ALBRECHT, A. M. & HUTCHISON, D. J. (1969). Folate reductase and specific dihydrofolate reductase activities of the amethopterin-sensitive *Streptococcus faecium* var. *durans*. *J. Bact.*, **100**, 533.

ANDERSON, B. A., PEART, M. B. & FULFORD-JONES, C. E. (1970). The measurement of serum pyridoxal by a microbiological assay using *Lactobacillus casei*. *J. Clin. Path.*, **23**, 232.

ANON. (1970). *Official Methods of Analysis*. 11th ed. (W. Horwitz, ed.), p. 782. Washington: AOAC.

BARTON-WRIGHT, E. C. (1962). *The Microbiological Assay of Vitamin B-Complex and Amino Acids*. London: United Trade Press. (See also *Lab. Pract.*, **10**, 543).

BIRD, O. D. & McGLOHON, V. M. (1972). Differential assays of folic acid in animal tissues. In *Analytical Microbiology*, vol. 2 (F. Kavanagh, ed.), p. 409. New York and London: Academic Press.

BOLINDER, A. E. (1972). Large plate assays for amino acids. In *Analytical Microbiology*, vol. 2 (F. Kavanagh, ed.), p. 479. New York and London: Academic Press.

150   L. B. PERRY, I. J. BOUSFIELD AND J. M. SHEWAN

88888888888888

BOUSFIELD, I. J. & MACKENZIE, A. R. (1972). The routine control of contamination in a culture collection. In *Safety in Microbiology* (D. A. Shapton and R. G. Board, eds), p. 73. London and New York: Academic Press.

CHANARIN, I., ANDERSON, B. B. & MOLLIN, D. L. (1958). The absorption of folic acid. *Br. J. Haesmat.*, **4**, 156.

COULTAS, M. K., ALBRECHT, A. M. & HUTCHISON, D. J. (1966). Strain variation within *Streptococcus faecium* var. *durans*. *J. Bact.*, **92**, 516.

COX, C. S. (1968). Method for the routine preservation of micro-organisms. *Nature, Lond.*, **220**, 1139.

DE MAN, J. C., ROGOSA, M. & SHARPE, M. E. (1960). A medium for the cultivation of lactobacilli. *J. appl. Bact.*, **23**, 130.

EIGEN, E. & SHOCKMAN, G. D. (1963). The folic acid group. p. 431 in *Analytical Microbiology* (F. Kavanagh, ed.), New York and London: Academic Press.

FLEMING, A. F., COMLEY, L. & STENHOUSE, N. S. (1971). Assay of serum and whole blood folate by a modified aseptic addition technique. *Am. J. clin. Nutrit.*, **24**, 1257.

FRY, R. M. (1954). The preservation of bacteria. p. 251 in *Biological Applications of Freezing and Drying* (R. J. C. Harris, ed.), New York: Academic Press.

FRY, R. M. & GREAVES, R. I. N. (1951). The survival of bacteria during and after drying. *J. Hyg., Camb.*, **49**, 220.

GARE, L. (1968). The use of depleted cells as inocula in vitamin assays. *Analyst, Lond.*, **93**, 456.

GIRDWOOD, R. H. (1963). The microbiological assay of vitamins and amino acids. *J. med. Lab. Technol.*, **20**, 26.

GORIN, N., MEULENHOFF, E. J. S. & YARROW, D. (1970). Determination of niacin in orange juice with lyophilized *Lactobacillus arabinosus* ATCC 8014. *Appl. Microbiol.*, **20**, 641.

GREENHAM, L. W. (1967). Letter to Editor on hazards of liquid nitrogen storage of glass containers. *J. med. Lab. Technol.*, **24**, 228.

GROSSOWICZ, N., MANDELBAUM-SHAVIT, F., DAVIDOFF, R. & ARONOVITCH, J. (1962). Microbiologic determination of folic acid derivatives in blood. *Blood*, **20**, 609.

GYÖRGY, P. & PEARSON, W. N. (ed.). (1967). *The Vitamins* (2nd ed., vol. 7). New York: Academic Press.

HANSEN, H. A. (1964). *On the Diagnosis of Folic Acid Deficiency*, p. 13. Stockholm: Almqvist & Wiksell.

HARTLEY, A. W., WARD, L. D. & CARPENTER, K. J. (1965). Hydroxylysine as a growth stimulant in microbiological assays for lysine. *Analyst, Lond.*, **90**, 600.

HOLLANDER, D. H. & NELL, E. E. (1954). Improved preservation of *Treponema pallidum* and other bacteria by freezing with glycerol. *Appl. Microbiol.*, **2**, 164.

HOWARD, D. H. (1956). The preservation of bacteria by freezing in glycerol broth. *J. Bact.*, **71**, 625.

KAVANAGH, F. (1963a). Elements of photometric assaying. p. 141 in *Analytical Microbiology* (F. Kavanagh, ed.). New York and London: Academic Press.

KAVANAGH, F. (ed.). (1963b). *Analytical Microbiology*, New York and London: Academic Press.

KAVANAGH, F. (ed.). (1972). *Analytical Microbiology*, vol. 2. New York and London: Academic Press.

KODICEK, E. & PEPPER, C. R. (1948). A critical study of factors influencing the microbiological assay of nicotinic acid. *J. gen. Microbiol.*, **2**, 292.

LAWRENCE, J. M., HERRINGTON, B. L. & MAYNARD, L. A. (1946). The nicotinic acid, biotin, and pantothenic acid content of cows' milk. *J. Nutrit.*, **32**, 73.

LEES, K. A. & TOOTHILL, J. P. R. (1955). Microbiological assay on large plates. 1. General considerations with particular reference to routine assay. *Analyst, Lond.*, **80**, 95.

LOY, H. W. & WRIGHT, W. W. (1959). Microbiological assay of amino acids, vitamins, and antibiotics. *Analyt. Chem.*, **31**, 971.

MANDERSON, G. J. & DOELLE, H. W. (1972). The effect of oxygen and pH on the glucose metabolism of *Lactobacillus casei* var. *rhamnosus* ATCC 7469. *Antonie van Leeuwenhoek*, **38**, 223.

MARTIN, S. M. & SKERMAN, V. B. D. (eds) with JONES, M. L. & QUADLING, C. (1972). *World Directory of Collections of Cultures of Microorganisms*, New York: Wiley-Interscience.

MEYNELL, G. G. & MEYNELL, E. (1965). p. 35 in *Theory and Practice in Experimental Bacteriology*. Cambridge: University Press.

NYMON, M. C. & GORTNER, W. A. (1946). Some culture studies on *Lactobacillus arabinosus* and *Lactobacillus casei*. *J. biol. Chem.*, **163**, 277.

OBERZILL, W. & GEISLER, CH. (1970). Einfluss methodischer Varianten auf die Standardkurve bei Vitamin B_{12}—Bestimmungen mit *Lactobacillus leichmannii*. *Path. Microbiol.*, **36**, 193 and 340.

OHARA, O. & SILBER, R. (1969). Studies on the regulation of one-carbon metabolism. The effects of folate concentration in the growth medium on the activity of three folate-dependent enzymes in *Lactobacillus casei*. *J. biol. Chem.*, **244**, 1988.

PEARSON, W. N. (1967). Thiamine. In *The Vitamins*. 2nd ed. (P. György & W. N. Pearson, eds). Vol. 7, p. 53 (see p. 70). New York: Academic Press.

PRICE, S. A. (1967). Assay of vitamins and amino acids. p. 55 in *Progress in Microbiological Techniques* (C. H. Collins, ed.). London: Butterworths.

RAE, P. G. & ROBB, P. M. (1970). Megaloblastic anaemia of pregnancy: a clinical and laboratory study with particular reference to the total and labile serum levels. *J. clin. Path.*, **23**, 379.

SCHOLES, P. M. (1961). The maintenance of microbiological assay organisms as freeze-dried cultures. *Analyst, Lond.*, **86**, 714.

SHEKLETON, M. C. & HAYNES, W. C. (1959). Microbiological assays with several strains of *Leuconostoc mesenteroides*. *J. Bact.*, **77**, 114.

SHOCKMAN, G. D. (1963). Amino acids. p. 567 in *Analytical Microbiology* (F. Kavanagh, ed.). New York and London: Academic Press.

SKEGGS, H. R. (1961). Current aspects of the vitamin B_{12} assay. *Devs ind. Microbiol.*, **2**, 159.

SKEGGS, H. R. (1963a). Biotin. p. 421 in *Analytical Microbiology* (F. Kavanagh, ed.). New York and London: Academic Press.

SKEGGS, H. R. (1963b). *Lactobacillus leichmannii* assay for vitamin B_{12}. p. 551 in *Analytical Microbiology*. (F. Kavanagh, ed) New York and London: Academic Press.

SKEGGS, H. R. (1967). Vitamin B_{12}. P. 277 in *The Vitamins*. 2nd ed. (P. György & W. N. Pearson, eds). Vol. 7. New York: Academic Press.

SMILLIE, R. (1967). The estimation of vitamin B_{12} and folic acid in biological material. Thesis, M.I.Biol. (Copy may be referred to in the members room of the Institute, 41 Queens Gate, London SW7).

SNELL, E. E. (1950). Microbiological methods in vitamin research. p. 372 in *Vitamin Methods* (P. György, ed.), vol. 1, New York: Academic Press.

SOKOLSKI, W. T., STAPERT, E. M. & FERRER, E. B. (1964). Liquid nitrogen freezing in microbiological assay. 1. Preservation of *Lactobacillus leichmannii* for direct use in the vitamin B_{12} assay. *Appl. Microbiol.*, **12**, 327.

SPRAY, G. H. (1964). Microbiological assay of folic acid activity in human serum. *J. clin. Path.*, **17**, 660.

STAMP, LORD, (1947). The preservation of bacteria by drying. *J. gen. Microbiol.*, **1**, 251.

STREETER, A. M. & O'NEILL, B. J. (1969). Effect of incubation time on the *L. casei* bioassay of folic acid in serum. *Blood*, **34**, 216.

TAMURA, T., SHIN, Y. S., WILLIAMS, M. A. & STOKSTAD, E. L. R. (1972). *Lactobacillus casei* response to pteroylpolyglutamates. *Analyt. Biochem.*, **49**, 517.

TANGUAY, A. E. (1959). Preservation of microbiological assay organisms by direct freezing. *Appl. Microbiol.*, **7**, 84.

TENNANT, G. B. & WITHEY, J. L. (1972). An assessment of work simplified procedures for the microbiological assay of serum vitamin B_{12} and serum folate. *Med. Lab. Technol.*, **29**, 171.

TOENNIES, G., FRANK, H. G. & GALLANT, D. L. (1952). Single system for measuring growth responses of three organisms to folacin and related factors (Bacterimetric studies. VII.). *Growth*, **16**, 287.

TOENNIES, G., WINEGARD, H. M. & GALLANT, D. L. (1950). Inhibition of growth of *Lactobacillus casei* by methionine and its relation to folic acid assimilation. *Arch. Biochem.*, **25**, 246.

WATERS, A. H. & MOLLIN, D. L. (1963). Observations on the metabolism in folic acid in pernicious anaemia. *Br. J. Haemat.*, **9**, 319.

A Simple, Rapid Assay for the Measurement of Antibiotic Concentrations in Human Serum

P. NOONE*, J. R. PATTISON AND R. C. B. SLACK

*Department of Bacteriology, School of Pathology,
Middlesex Hospital Medical School,
Riding House Street, London W1P 7LD, England*

Conventional methods for the measurement of concentrations of anti-biotics in serum are based on the inhibition of growth over 16–18 h of a test organism by dilutions of test serum using either a tube dilution technique or an agar diffusion method, comparing growth inhibition with that produced by standard concentrations of antibiotic (Garrod and O'Grady, 1971). An 18 h delay in obtaining the result is acceptable only in certain situations, for example in demonstrating adequate therapeutic levels of penicillin during a 4–6 week course for endocarditis. However, in circumstances when a potentially toxic antibiotic is being used in seriously ill patients, this delay is a serious drawback. For example, in the treatment of Gram negative septicaemia with gentamicin, accumulated experience indicates that approximately 80% of causative organisms are sensitive to a concentration of 4 mcg/ml, whereas concentrations of 10 mcg/ml or greater have been found in approximately 60% of the patients with possible ototoxicity in whom serum concentrations have been measured (Arcieri *et al.*, 1970). It was therefore considered neces-sary to monitor serum gentamicin concentrations in order to avoid toxic over-dosage particularly in patients with impaired renal function. During the 18 h delay in achieving the result by a conventional assay, the situation can alter dramatically as three more doses of the antibiotic may have been given and renal function may have altered. It was with these problems in mind that an attempt was made to develop a rapid assay technique. Having used such methods for over two years it is now clear that, with respect to gentamicin, their more important contribution is in providing a rational basis for increasing the recommended dosage in many patients in order to

* Present address: Bacteriology Department, Royal Free Hospital, London WC1, England.

achieve therapeutic serum concentrations rather than in decreasing the dosage to avoid toxicity.

Theoretical Considerations

In 1968 Faine and Knight described a rapid assay method based on the influence of antibiotics on the breakdown of lactose by *Klebsiella* spp as measured by the fall in pH of the medium. Although the detailed method they described was cumbersome and difficult to use, the theoretical basis of the assay was strikingly simple and in practical terms seemed only to require accurate measurement of pH. With this in mind, the possibility of utilizing a system (Noone, Pattison and Samson, 1971) involving the hydrolysis of urea by *Proteus* spp was investigated as this reaction is obviously rapid and the pH changes in an alkaline direction which favours the antimicrobial activity of gentamicin, and all aminoglycosides.

Urea hydrolysis

Figure 1 illustrates the rise in pH of a 2% (w/v) urea medium inoculated with *Proteus mirabilis* NCIB 10823 showing a lag period of *c*. 20 min and

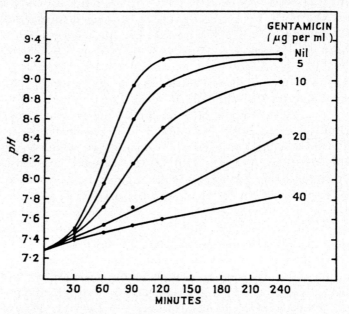

Fig. 1. Rate of change of pH.
5 ml urea medium plus 0·4 ml human pool serum and 0·4 ml phosphate buffer (gentamicin concentration as indicated) inoculated with *P. mirabilis* and incubated at 37°.

a period of rapid pH change which tails off after incubation for 2 h. Such an extensive pH change after 90–120 min incubation is dependent entirely upon the presence in the medium of readily available amino-acids (peptones or "casamino-acids"), but is independent of glucose. In contrast the multiplication of a heavy inoculum $(4-8 \times 10^7$ organisms/ml) of *Pr. mirabilis* in 2% (w/v) urea broth shows a lag period of 60 min and a first doubling-time at 120 min. The maximum rate of multiplication of *Pr. mirabilis* in the presence of 2% (w/v) urea is dependent on glucose but is independent of peptone. Up to 120 min incubation, urea hydrolysis is thus largely independent of bacterial multiplication in the system described. If urea hydrolysis is measured as ammonia produced, using the Berthelot alkaline hypochlorite reaction, a lag period of 15–20 min is again found indicating that this initial delay before the rise in pH is not due to buffering by the medium. All these observations with intact organisms suggest that the urease of *Pr. mirabilis* is an inducible enzyme. Experiments with cell-free extracts indicate that urease activity is present only in very low concentrations when the organism is grown in the absence of urea. Growth in the presence of urea, however, induces intense urease activity after a lag period of *c.* 15 min. This is in agreement with other observations (D. S. Bascomb, pers comm) that the ureases of *Pr. mirabilis* and *Proteus vulgaris* are inducible enzymes whereas those of *Proteus morganii* and *Proteus rettgeri* are constitutive enzymes.

The effect of antibiotics

Figures 1 and 2 illustrate the progressive inhibition by increasing concentrations of gentamicin of the pH change of 2% (w/v) urea broth inoculated with *Pr. mirabilis*. The minimum inhibitory concentration (MIC) of gentamicin for this organism is 0·6 mcg/ml and a differential, inhibitory effect of the pH change of the medium can be detected between the 1·25 and 160 mcg/ml gentamicin standards, i.e. final concentration 0·09 to 11·4 mcg/ml (Fig. 2). The test organism *Pr. mirabilis* was chosen initially because of its resistance to many of the commonly used antibiotics, a property which was thought to be potentially useful for the assay of gentamicin in the presence of other antibiotics. On routine disc testing (Oxoid Multidisk), the organism is sensitive to gentamicin (10 mcg), kanamycin (5 mcg), neomycin (10 mcg) and chloramphenicol (30 mcg) but resistant to streptomycin (10 mcg), ampicillin (10 mcg), tetracycline (10 mcg), cephaloridine (5 mcg), cotrimoxazole (25 mcg), carbenicillin (100 mcg), polymyxin B (300 units), penicillin (2 mcg), novobiocin (5 mcg) and rifamide (10 mcg). With this organism, inhibition of the rise in pH is found with kanamycin over the range 0·1–7·0 mcg/ml and chloramphenicol (range 3–10 mcg/ml). Using different test organisms sensitive to the appropriate antibiotic a similar inhibition is found with streptomycin (0·75–15 mcg/ml) and tetracycline (1·6–50 mcg/ml).

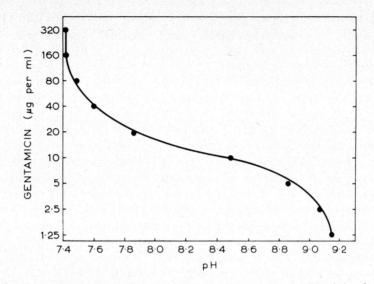

FIG. 2. Inhibition of pH change by gentamicin Reaction mixture as Fig. 1, incubated for 90 min.
5 ml urea medium plus 0·4 ml human pool serum and 0·4 ml phosphate buffer (gentamicin concentration as indicated) inoculated with *P. mirabilis* and incubated at 37°.

However, even with test organisms sensitive to the antibiotics, no inhibitory effect of the potential pH change after 90–120 min incubation is found with ampicillin (up to 200 mcg/ml), carbenicillin (up to 500 mcg/ml) cephaloridine (up to 400 mcg/ml), trimethoprim (up to 50 mcg/ml), sulphafurazole (up to 500 mcg/ml) and trimethoprim/sulphafurzaole combinations (up to 25:500 mcg/ml combination). This has important practical implications since no pre-treatment of test solutions in order to inactivate penicillins or cephalosporins prior to assaying for gentamicin will be necessary. It is noted that all the antibiotics which affect the assay act by inhibiting protein synthesis. With gentamicin at concentrations below the MIC for the test organism, the inhibition of urease activity is presumed to be due to inhibition of the synthesis of the enzyme itself. Gentamicin is rapidly bactericidal and, at concentrations above the MIC, the inhibition of the pH change after 90 min incubation in 2% (w/v) urea broth correlates well with the percentage of bacteria killed in the first 20 min, i.e. in the induction time of the enzyme. It does not correlate, however, with the bacteria killed at 90 min indicating that, once formed, the urease will continue to split urea irrespective of the fate of the bacterial cells. Since there is a lag period of 60 min prior to bacterial multiplication in 2% (w/v) urea broth the antibiotics which act by inhibition of cell-wall synthesis are not bactericidal before urease induction. Therefore, they have no effect on the change of pH of the medium during the assay.

Effect of sodium chloride

The range of gentamicin concentrations over which a differential inhibition of the rise in pH is found is decreased approximately ten-fold if sodium chloride is omitted from the medium and this alteration of sensitivity is independent of the presence of glucose or peptone but can be produced by changes in potassium chloride concentration of corresponding ionic strength. The increase in sensitivity is paralled by an increase in the bactericidal effect of gentamicin for *Pr. mirabilis* at low salt concentrations. This finding is in agreement with Rubenis, Kozij and Jackson (1964) and Medeiros, O'Brien, Wacker and Yulug (1971) who demonstrated a rising MIC of gentamicin for *Pr. mirabilis*, *Escherichia coli*, *Pseudomonas aeruginosa* and *Staphylococcus aureus* with increasing sodium chloride concentrations in the medium.

Practical Details

Organism

Proteus mirabilis NCIB 10823 is maintained by daily subculture in 0·2% w/v) glucose broth (Todd Hewitt) buffered to pH 7·8 and onto C.L.E.D. agar for purity. The broth is inoculated each day from the purity plate. The inoculum used in the test is taken from the overnight broth culture, well mixed but not further standardized. With 0·1 ml of the suspension added, a viable count of $4·0 \pm 0·4 \times 10^7$ organisms/ml is obtained.

Media

The initial development of the assay was based on the use of commercially available urea broth base medium (Oxoid CM71) adjusted to pH 7·1. The base medium is stored in 90 ml amounts after autoclaving (121°/10 min) and 10 ml of 40% (w/v) urea, sterilized by Seitz filtration, is added prior to use. The complete urea medium is then dispensed in 3 or 5 ml amounts in sterile disposable plastic bottles making sure that in any given test all the medium comes from the same stock bottle. Experiments have shown that exactly similar results can be obtained with medium made up from the declared constituents of Oxoid CM 71.

For those procedures which involve the use of medium without added NaCl, the salt free urea broth consists of 4% (w/v) urea in 0·01 M phosphate buffer pH 7·0 containing 0·1% (w/v) peptone and 0·0004% (w/v) phenol red.

Gentamicin standards

Gentamicin solutions dispensed for therapeutic use have been found to be a satisfactory source of standard. A stock solution of 400 mcg gentamicin/ml is prepared by taking 1 ml from a commercial vial containing 80 mg in 2 ml and diluting in 100 ml distilled water. This may be kept for at least 2 months at 4°.Ten ml quantitites of standards of concentrations 40, 20, 10, 5 and 2·5 mcg/ml are made up weekly by diluting from the stock solution with 0·1 M phosphate buffer (pH 8·0). In the assay described below, no direct comparison of the inhibitory effect of the test serum with a standard curve is made; the accuracy of the concentration of standard solutions is not critical within 10%, but it is vital that they are accurate doubling-dilutions of the highest concentration.

Test procedures

All attempts to devise a test based on the inhibitory effect of a sample of the test serum compared with the inhibition produced by the same amount of pooled human serum containing standard gentamicin concentrations have been unsuccessful. This is due to variations in individual sera of many factors which may affect the system and are not fully understood as yet. If, however, two standard curves are developed which are identical except that one has a sample of neat test serum and the other has half this amount, then any difference in pH is proportional to half the gentamicin concentration in the test serum. All the methods described below depend on this convention.

Serum

Antibiotic-free, pooled human serum, after Millipore filtration, is stored at −20° in aliquots which are thawed for use.

Equipment

In practice we use 10 ml pipettes for media and Eppendorff-type pipettes for the sera and standards except in the micro method when a microtitre dropper (Cooke Engineering Co. Ltd.) delivering 0·025 ml may be used. For bulk dispensing, a Hook and Tucker dilutor has been used with good results but it is only justified where at least 10 assays are performed weekly. A pH meter reading to 0·01 units is required—e.g. Pye model 290 fitted with a Pye Ingold E07 electrode.

Incubation

This can vary with the gentamicin concentration to be measured, but the usual time of incubation in a 37° water-bath is 75–90 min. The time to read the pH is when the phenol red indicator has turned red in the bottles with low concentrations but remains relatively unchanged in the bottles with the highest concentration (Fig. 3).

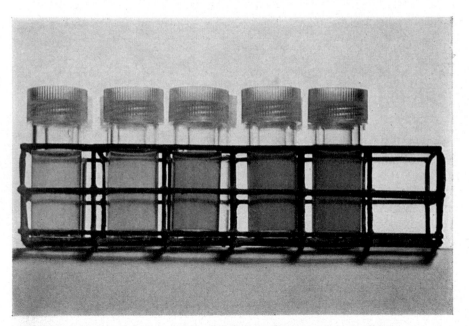

FIG. 3. Change in indicator (phenol red) in medium containing antibiotics.

Method

Label and set up two rows of 5 bottles with the following additions:

First row

Bottle Label	40+x	20+x	10+x	5+x	2·5+x
3 ml Oxoid CM71					
4% urea	+	+	+	+	+
0·2 ml Gentamicin standards of concentration (mcg/ml)	40	20	10	5	2·5
0·2 ml Test serum	+	+	+	+	+

Second row

Bottle Label	$40+^x/2$	$20+^x/2$	$10+^x/2$	$5+^x/2$	$2·5+^x/2$
3 ml Oxoid CM71					
4% urea	+	+	+	+	+
0·2 ml Gentamicin standards of concentration (mcg/ml)	40	20	10	5	2·5
0·1 ml Test serum	+	+	+	+	+
0·1 ml Pool serum	+	+	+	+	+

Inoculate all bottles with 0·1 ml of a mixed, overnight broth culture of *Pr. mirabilis*. Mix well by inversion and incubate at 37° in a water-bath.

Reading

When the colour change is judged sufficient the pH is read, after mixing by inversion, in the order $40+x$, $40+^x/2$, $20+x$, $20+^x/2$ etc as rapidly as possible. This order is important as the rate of change of pH at this stage of the reaction is rapid and the important readings are the differences between x and $^x/2$ for each standard.

Results

As an example the following pattern of results might be obtained:

Bottle label	pH	Bottle label	pH
40+x	7·42	$40+^x/2$	7·44
20+x	7·69	$20+^x/2$	7·82
10+x	8·08	$10+^x/2$	8·26
5+x	8·24	$5+^x/2$	8·42
2·5+x	8·34	$2·5+^x/2$	8·50

Plot these results on semi-logarithmic graph paper with pH along the abscissa and gentamicin standard concentration (i.e. 40, 20, 10, 5, 2·5) along the ordinate (Fig. 4). In this manner, the vertical axis does not give the actual final gentamicin concentration but the concentration of the original standard itself. This method of plotting the results avoids difficulties in calculations to derive x. Two curves (A and B) are drawn through the points obtained for each row of bottles and the vertical difference at each 0·05 or 0·1 pH unit measured. The average of these values multiplied by 2 gives the gentamicin concentration in the test serum.

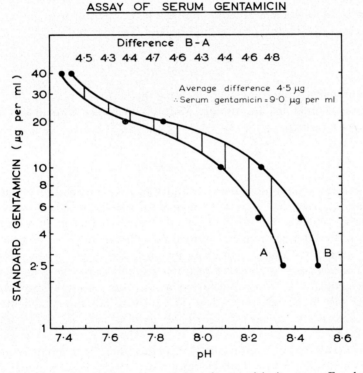

FIG. 4. Method for determining concentration of gentamicin in serum. For details of test, see text.

Micro-method

The previous method requires 1·5 ml of serum for each assay. For paediatric work this is not acceptable, so the following procedure has been devised to reduce serum requirements to less than 0·4 ml. The method

is identical to that described previously except that 5 ml of salt-free 4%
urea medium, 0·05 ml gentamicin standards, 0·05 or 0·025 ml test serum
with 0·025 fl pooled serum are added to appropriate bottles. Accurate
dispensing of these volumes is so essential that a micro-titre dropper is
required, or a suitable Eppendorff-type pipette.

Low concentrations

For gentamicin concentrations below 2 mcg/ml the above methods are
relatively inaccurate. There is little occasion in clinical practice to require
an estimation in this range, but, if it is necessary, the same volumes may
be used as in the standard method but using salt-free 4% urea medium with
gentamicin standards 5, 2·5, 1·25, 0·63 and 0·31 mcg/ml.

High concentrations

For gentamicin values above 15 mcg/ml the accuracy is again decreased.
For these estimations appropriate dilutions in pooled human serum give
satisfactory results.

Quality control

With serum containing known amounts of gentamicin, the precision is
within 15% over the range 3–15 mcg/ml using the standard assays and,
using the methods described above, the same coefficient of variation is
found in the extended range 1–30 mcg/ml. This includes all clinical situations. Generally, there is a trend to produce low results when the gentamicin concentration is above 15 mcg/ml. Initially cross-checking of results
with a standard 18 h agar diffusion assay using a spore suspension of
Bacillus subtilis carried out both in the authors' laboratory and indepen-
dently showed the urease assay to be at least as accurate as this method.
Reproducibility of the urease assay is such that it is now no longer considered necessary to perform duplicates since there is inherent replication
in a single assay (Noone, Pattison and Slack, 1973). Pooled human serum
to which a known amount of gentamicin has been added is run as a quality
control.

Errors

Errors which cannot be entirely accounted for by technique rarely occur
and the most likely source is concurrent administration of chloramphenicol
or kanamycin. The buffering of the medium itself plus the additional

buffering of the standards will account for variations in sera, but in early experiments severely acidotic or uraemic patients gave small but consistent inaccuracies (Froud, 1971). The latter is compensated for by increasing the urea concentration from 2% as originally described to 4% (w/v) in the medium (Noone, Pattison and Slack, 1972). Errors giving "wild" results in single bottles can be seen as the curves will emphasize a single inaccuracy. At both extremes of the curves there are often inaccuracies, so these parts are usually excluded when deriving the mean. This is because at the higher concentrations, the error of reading off the logarithmic scale is too great and at the 2·5 gentamicin standards the rate of change of pH slows down above pH 8·7. But in between these values there is an almost a linear relationship of pH change to log gentamicin concentration. If there is insufficient serum to use all points, the top standards $40+x$, $40+^x/2$ and the lowest $2·5+^x/2$ may be omitted without altering the derivation of the curves.

Estimation of other antibiotics

The original method has been adapted successfully to estimating serum concentrations of kanamycin, streptomycin, chloramphenicol and tetracycline. For kanamycin, the concentrations of the standards are 100, 50, 25, 12·5, 6·25 mcg/ml using *Pr. mirabilis* NCIB 10823 as the test organism. This organism can also be used for chloramphenicol assay with standards 50, 25, 12·5, 6·25 and 3·1 mcg/ml. For streptomycin assays a test organism, *Pr. mirabilis* BS 712, sensitive to this antibiotic, is used with standards of 200, 100, 50, 25 and 12·5 mcg/ml and for tetracycline a sensitive *Pr. vulgaris* is used with standards of 50, 25, 12·5, 6·25 and 3·1 mcg/ml. For tobramycin, *Pr. mirabilis* NCIB 10823 is used as test organism with standards of 20, 10, 5, 2·5 and 1·25 mcg/ml using a salt-free broth as in the micromethod described above. Tobramycin is only as active as gentamicin against *Pr. mirabilis* (MIC 0·6 mcg/ml).

Assaying other fluids

Using the original method to try to assay gentamicin in urine, we obtained wide variation in results but using the micromethod accuracy was improved. By using the corresponding pooled blank fluid, the method can be adapted for measuring gentamicin or aminoglycosides in urine, CSF, pleural and peritoneal fluids. However, there is not often much clinical demand in such estimations, so the conventional 18 h agar diffusion method may be preferred.

Monitoring Gentamicin Therapy

Gentamicin is now the antimicrobial agent of choice for severe Gram-negative sepsis (Martin *et al*, 1969). It is essential that adequate serum concentrations are reached as soon as possible (Noone *et al.*, 1974). For this reason alone, it is necessary to employ a rapid assay. In addition, when monitoring for possible toxicity which may be due to one or a combination of the following factors of high concentrations, prolonged therapy or accumulation of the drug (Arcieri *et al.*, 1970) and Jackson and Arcieri (1971), frequent sera need to be assayed, especially in patients with renal impairment.

When to take blood for assay

1. "Peak concentrations" *c.* 1 h after intra-muscular or 10–15 min after intravenous injection.
2. "Trough concentrations" just before next dose—usually 8 h after previous administration.

If possible "peak bloods" should be taken after the first injection and whenever the dosage is adjusted, or there is a change in renal function, judged by rising blood urea or alteration in urine output. "Trough sera" should be taken in all cases of renal impairment or when accumulation may be suspected as in a prolonged course of therapy. They should also be taken during or after dialysis.

Dosage

For adults, at least 5 mg of gentamicin/kgm body weight in 3 divided doses must be given daily. This would correspond to 120 mg three times-a-day for a 70 kgm adult. Lower doses will give adequate urine concentrations in patients with urinary tract infections. The dose should be increased if serum concentrations greater than 5 mcg/ml are not achieved after the first injection. If values above 12 mcg/ml are obtained, it may be considered necessary to reduce the dose but, in patients with serious sepsis, the risks of inadequate treatment far outweigh those of possible toxicity. In cases of bacteraemia, pneumonia and peritonitis peak concentrations in the range 8–12 mcg/ml should be achieved (Noone *et al.*, 1974).

With infants and especially neonates, larger doses need to be given. In our limited experience adequate serum concentrations are only achieved with doses of 7·5 mg/kgm/day and more. This is in agreement with

McCracken, Chrane and Thomas (1971) who recommend this dose for infants over one week in age.

Comparison of methods

Methods which attempted to adjust gentamicin dosage by measurement of blood urea (Gingell and Waterworth, 1968), creatinine clearance (McHenry *et al.*, 1971) or by complicated calculation (O'Grady, Brown, Gaya and Mackintosh, 1971) have obvious disadvantages over a direct serum antibiotic assay, particularly when this can be related to the last dose. For rapid assays, routine laboratories at present use either the urease or agar diffusion methods (Sabath *et al.*, 1971; Winters, Litwack and Hewitt, 1971; Reeves, 1972).

Chemical assays are available and obviously offer better precision but require liquid scintillation counters. Smith, Van Otto and Smith (1972) have devised a method depending on the adenylation of gentamicin by a specific enzyme and its binding on to phosphocellulose paper. In this system labelled ATP is used, which is easier and cheaper to obtain than radioactive gentamicin which is required for the radioimmunoassay of Lewis, Nelson and Elder (1972). Although they yield results within 4 h,

TABLE 1.

Speed	Urease Within 2 h	Agar diffusion Minimum 4 h
Operator time		
setting up batch	1 h/week	1 h/week
individual assay	5 min	15 min
reading/calculation	10 min	10 min
subculturing organism	1 min/day	1 min/day
Cost		
capital	pH meter/pipettes	plates ($\pm$ automated equipment)
per test	25p approximately	15p approximately
Sensitivity	1–30 mcg/ml	1–30 mcg/ml
Specificity	Good except chloramphenicol kanamycin measures antimicrobial activity	varies with test organism, can be non-specific
Precision		
average error	8%	11%
range of error	0–20% (Noone, Pattison and Garfield Davies 1974)	usually 1–16% (Sabath *et al.*, 1971)
Pre-treatment of sera	Not required	Often necessary

both these methods require extensive methodology, the expertise and expense of which is not within the reach of most clinical laboratories. Thus the only relevant comparison is between the urease and agar diffusion methods (Table 1). The urease method has been statistically validated (Noone, Pattison and Garfield Davies, 1974), for accuracy and reproducibility.

The chief advantages of the urease assay are its speed and freedom from interference by most other antibiotics which can produce false high values with an agar diffusion method. In the early stages of treatment of patients with severe sepsis, both these are critical.

Conclusions

The urease assay for aminoglycoside antibiotics in serum can give results within 2 h. It does not require expensive media or equipment and is technically simple. Results are reproducible within 15 % which is acceptable for clinical use. There is a need for closer monitoring of all seriously ill patients on antibiotics, and the method described enables laboratories to assay gentamicin rapidly thus providing a rational basis for therapy.

References

ARCIERI, G. M., FALCO, F. G. & SMITH, H. M. (1970. Clinical research experience with gentamicin: Incidence of adverse reactions. *Med. J. Aust.*, **1**, 30.
FAINE, S. & KNIGHT, D. C. (1968). Rapid microbiological assay of antibiotic in blood and other body fluids. *Lancet*, **ii**, 1425.
FROUD, D. J. R. (1971). Serum gentamicin levels. *Lancet*, **ii**, 1425.
GARROD, L. P. & O'GRADY, F. (1971). *Antibiotic and chemotherapy*. Livingstone. Edinburgh, p. 475.
GINGELL, J. C. & WATERWORTH, P. M. (1968). Dose of gentamicin in patients with normal renal function and renal impairment. *Brit. Med. J.*, **2**, 19.
JACKSON, G. G. & ARCIERI, G. M. (1971). Ototoxity of gentamicin in man. *J. Infect. Dis.*, **124**, Suppl. 130.
LEWIS, J. E., NELSON, J. C. & ELDER, H. A. (1972). Radioimmunoassay of an antibiotic: gentamicin. *Nature, New Biol.*, **239**, 214.
MARTIN, C. M., CUOMO, A. J., GERAGHTY, M. J., ZAGER, J. R. & MANDES, T. C. (1969). Gram-negative rod bacteraemia. *J. Infect. Dis.*, **119**, 506.
McCRACKEN, G. H., CHRANE, D. F. & THOMAS, M. L. (1971). Pharamcologic Evaluation of gentamicin in newborn infants. *J. Infect. Dis.*, **124**, Suppl. 214.
McHENRY, M. C., GAVAN, T. L., GIFFORD, R. W. et al. (1971). Gentamicin dosages for reneal insufficiency. *Ann. Int. Med.*, **74**, 192.
MEDEIROS, A. A., O'BRIEN, T. F., WACKER, W. E. C. & YULUNG, N. F. (1971). Effect of salt concentration on the apparent *in-vitro* susceptibility of pseudomonas and other gram-negative bacilli to gentamicin. *J. Infect. Dis.*, **124**, Suppl. 59.
NOONE, P., PATTISON, J. R. & SAMSON, D. (1971). Simple, rapid method for assay of aminoglycoside antibiotics. *Lancet*, **ii**, 16.

NOONE, P., PARSONS, T. M. C., PATTISON, J. R., SLACK, R. C. B., GARFIELD
DAVIES, D. & HUGHES, K. (1974). Experience in monitoring gentamicin
therapy during treatment of serious gram-negative sepsis. *B.M.J.*, **1**, 477.

NOONE, P., PATTISON, J. R. & GARFIELD DAVIES, D. (1974). The effective
use of gentamicin in life threatening sepsis. *Postgrad. Med. J.* (in press).

NOONE, P., PATTISON, J. R. & SLACK, R. C. B. (1972). Rapid assay of gentamicin.
Lancet, **ii**, 1194.

NOONE, P., PATTISON, J. R. & SLACK, R. C. B. (1973). Assay of gentamicin.
Lancet, **i**, 49.

O'GRADY, F., BROWN, W. R. L., GAYA, H. & MACKINTOSH, I. P. (1971).
Antibiotic levels on continuous intravenous infusion. *Lancet*, **ii**, 209.

REEVES, D. S. (1972). Assay of gentamicin. *Lancet*, **ii**, 1369.

RUBENIS, M., KOZIJ, V. M., & JACKSON, G. G. (1963). Laboratory studies on
gentamicin. p. 153 in Antimicrobial Agents & Chemotherapy.

SABATH, L. D., CASEY, L. J., RUCH, P. A., STUMPF, L. L. & FINLAND, M.
(1971). Rapid microassay of gentamicin, kanamycin, neomycin, streptomycon
and vancomycin in serum or plasma. *J. Lab. Clin. Med.*, **78**, 457.

SMITH, D. H., VAN OTTO, B. & SMITH, A. L. (1972). A rapid chemical assay
for gentamicin. *New Eng. J. Med.*, **286**, 583.

WINTERS, R. E., LITWACK, K. D. & HEWITT, W. L. (1971). Relation between
dose and levels of gentamicin in blood. *J. Infect. Dis.*, **124**, Suppl. 90.

Bio-assay of Agricultural and Horticultural Fungicides

E. C. Hislop and D. R. Clifford

*University of Bristol, Long Ashton Research Station,
Long Ashton, Bristol BS18 9AF, England*

Preface

Although it is hard to find up-to-date assessments of the economic importance of fungicide usage in the literature, Cramer (1967) has suggested that
even when they were used the annual losses due to fungus diseases amounted to some 500 m tons of food, worth about £1,000 million. It is very
difficult to quantify what is meant by loss in this context, although attempts
to do so have been made by Ordish (1952), Ordish and Mitchell (1967)
and Ordish and Dufour (1969). The total world usage of fungicides in
1965 was some 150 m pounds weight (Shepard, Mahan and Fowler, 1966)
and with an expanding world population having expectations of a higher
standard of living and greatly increased food consumption this figure is
bound to increase. However, in perspective, fungicides account only for
about 10% of all pesticides used.

New and improved fungicides result from research and development
programmes which may take 5–10 years (Anon, 1972) and involve $3–5
million (Wellman, 1967; Anon, 1972) for each material which finally
reaches the market. Wellman (1959) estimated that 5,000 compounds were
screened every year as potential agricultural pesticides but that com-
merically successful products amounted to only one per year. Current
data regarding the number of candidate compounds are not available but
we believe that at least a similar number of materials are screened, most of
which probably arise from industrial synthetic programmes, and there is
obviously a radical need for rapid, reliable, efficient and comprehensive
bio-assay methods. Furthermore, the discovery of fungicides better
than those currently available will become increasingly difficult and
require more efficient and sophisticated methods which are capable of
recognising novel modes of action.

Introduction

Bio-assay techniques suitable for the evaluation of potential agricultural fungicides fall broadly into two classes, viz tests *in vitro* carried out in the laboratory in the absence of a host plant, and tests *in vivo* where the chemical is applied to plants grown in the glasshouse or in the field.

Traditionally, the steps in the search for a new fungicide are laboratory assay, glasshouse evaluation and field testing in this order, but opinions vary amongst investigators as to the importance of each step. Whether or not this scheme is followed will depend upon the aims of the investigator, and 20 years ago less emphasis would have been placed on glasshouse testing (Horsfall, 1945; Horsfall and Rich, 1951; Horsfall, 1956). However, rigid adherence to this practice would have precluded for example the development of the important powdery mildew fungicide, dinocap (Rich and Horsfall, 1949). McCallan (1959) suggested that all of the new protectant fungicides developed during the previous 25 years had been discovered by means of laboratory tests. However, he also stated that while laboratory methods continued as a primary screen in many programmes in the U.S.A., in Europe and elsewhere there was an increasing interest in developing the less precise glasshouse methods of evaluation which he regarded as being more closely correlated with field performance. In fact, at least two major agro-chemical producers in the U.K. now attach very much more importance to routine glasshouse tests than to laboratory tests. In the specific case of bio-assay of soil fungicides, it has been established that a true evaluation is obtained only in direct tests using soil (Kendrick and Middleton, 1954; Zentmeyer, 1955; McCallan, 1959). A description of objectives and methods of testing fungicides in a Dutch Company has been presented by Tempel (1969).

Much of the information in the present paper has been abstracted from a large number of specialist publications. Our object here is to provide a unified practical guide to the principles and practices of bio-assay of chemicals destined to be used as agricultural and horticultural fungicides, although many of the principles involved are equally applicable to the evaluation of chemicals for other biological purposes.

Considerations of the use of fungicides in fields other than agriculture have been presented by Block (1967) for industrial fungicides and by Baechler (1967) for wood preservatives.

Selection of fungi for use in bio-assay

Although there are more than 1 million known fungi, Ordish and Mitchell (1967) state that fewer than 200 of these are plant pathogens while less than

10% of the pathogens cause serious losses. In theory, any fungus can be used for the investigation of the fungitoxic activity of chemicals *per se* and in this case the choice of the fungus is dictated by the type of assay to be used. For example, any fungus which produces spores which germinate well under the test conditions can be used for assays of inhibition of spore germination, but in practice most workers only use as test fungi those that grow easily in culture and have large spores which can be readily counted under a microscope. On the other hand, fungi which rarely produce spores may still be of use for the measurement of mycelial growth in the presence of test materials. If it is necessary to use fungi which cannot be cultured *in vitro*, spores must be obtained from infected host plants and their viability is often more variable than that of spores grown in artificial culture.

Within the above limitations it is desirable whenever possible to test potential agricultural fungicides against plant pathogenic fungi, and assays *in vivo* (where inhibition of lesion development is to be measured) necessitate the use of compatible host/pathogen combinations. The choice of such combinations is also influenced by a consideration of the characteristics of the host plant, as discussed more fully under "Glasshouse Tests". In commercial practice, the activity of a given compound against a range of pathogenic fungi is examined in both laboratory and glasshouse screening programmes. However, research laboratories are often committed either to studies on a particular pathogen or group of pathogens or to such fundamental investigations as mode of action at the cellular level, and here the choice of fungus is determined by the specific requirements of the investigation.

The evaluation of differential sensitivity of one fungus to a series of chemicals or a range of fungi to one chemical is of considerable importance. Classically, fungi were broadly divided into two groups, i.e. those susceptible or resistant to copper or to sulphur; for example, powdery mildews are susceptible to sulphur, whilst downy mildews are controlled by copper. However, the introduction of the more complex organic fungicides led to subtle differences in fungitoxic spectra: for instance, zinc dimethyldithiocarbamate (ziram) controls early blight of potato (caused by *Alternaria solani*) but not late blight (caused by *Phytophthora infestans*), whilst zinc ethylenebisdithiocarbamate (zineb) controls both with equal efficiency (Horsfall, 1956). Similar specificity for individual fungus/chemical systems was demonstrated by Casarini and Pucci (1957) who suggested that a potential fungicide should be tested in the laboratory against the same fungus as is responsible for the disease to be controlled in the field. In contrast however, McCallan, Wellman and Wilcoxon (1941) using spore germination tests involving 6 fungi and 20 compounds, and Neely and Himelick (1966) with 7 fungi and 24 compounds ranked the activities of the candidate fungicides in more or less the same order irrespective of the fungus used.

The influence of the host plant on fungicide activity should not be overlooked. Although Munroe (1962), working with ten foliage fungicides and two pathogens on a common host (broad bean), showed that for each disease all the fungicides were ranked consistently in the same order of fungitoxicity by tests *in vitro* and tests in the glasshouse and field. Marsh (1936), Curtis (1944), Yarwood (1945) and more recently Clifford and Hislop (1973) have shown that fungitoxic activities *in vivo* may differ significantly from those observed *in vitro*.

Many of the successful commercial fungicides sold today have a broad spectrum of activity, but the more recent introduction of systemic fungicides [(which often have highly selective modes of action, in some cases sometimes resulting in greater specificity of action) emphasizes the need for screens] involving a wide range of fungal species.

Evaluation of fungicides in the laboratory

Laboratory assay of potential fungicides has been reviewed by a number of investigators (Horsfall, 1945, 1956; McCallan, 1947, 1959; Torgeson, 1967; Neely, 1969) and the following represents a selection of the more important aspects of this field.

Fungicides may be evaluated in the laboratory in the presence or absence of plant material. In modern usage, the term *"in vitro"* is used to describe tests performed in or on such varying materials as glass, agar and cellulose film and usually in the absence of plant material: we confine the use of the term *"in vivo"* to experiments in which intact host plants, seeds or other plant organs are used. In some cases, distinction between these classifications becomes difficult as, for example, when materials are tested on detached leaves or leaf disks. The term "fungicidal" should only be used to describe the action of those materials which kill fungi, whilst those which merely prevent or delay growth are described as "fungistatic". The term "fungitoxic" is used in situations where a material causes morphological or physiological change but where no effort is made to determine whether or not the fungus is killed.

Fungi react to the presence of toxic chemicals in a variety of ways, the responses most commonly utilized for evaluation of fungitoxicity being suppression of growth, respiration and germination, and each of these are considered below.

Suppression of fungal growth in culture

This is one of the most simple and inexpensive techniques that has been evolved for the bio-assay of fungitoxicity and obviates the use of a microscope. Commonly, test chemicals are incorporated into a nutrient agar or a broth which is then seeded with fungal spores or mycelium, incubated under controlled conditions and the reduction in growth, compared with that on a control treatment, is estimated. This technique is invaluable for investigations with fungi which do not spore easily. There are many variations of this general method, but in the most simple form the fungus is grown on a plate of agar and the diameter of the colony is measured or, with broth techniques, linear growth or weight of mycelium is evaluated (Mason and Powell, 1947; Le Tourneau and Buer, 1961). Rather more specialized variations of the agar plate technique have been described by Forsberg (1949) and Manten, Klöpping and van der Kerk (1950). Sometimes, chemicals are applied to a particular area of an agar plate, the plate seeded with the fungus and zones of inhibition of fungal growth extending from the toxicant are measured. In other variations, the chemical may be placed in holes cut in the agar (Mildner *et al.*, 1963), or added to filter paper disks (Thornberry, 1950; Leben and Keitt, 1950) or pieces of string (Kuhfuss, 1957) which are then placed on top of the agar.

Materials destined to be used as potential fungitoxic agents in soil pose certain specific problems in that tests carried out in the laboratory in the absence of soil tend to give misleading information (McCallan, 1948; de Tempe, 1972). In order to avoid this problem, Pote and Thomas (1954) leached test formulations through a column of sterile soil before assaying the eluant in an agar plate test. Zentmeyer (1955), Munnecke (1958) and de Tempe (1972) have also evolved specialized techniques for the investigation of potential soil fungicides.

The agar plate assay technique is also valuable for the study of such factors as the role of chelation in fungitoxic activity (Byrde, Clifford and Woodcock, 1958, 1961) and antisporulant activity (Sneh, Clifford and Corke, 1972), although Lukens and Horsfall (1968) successfully used disks of filter paper for the latter purpose.

Results obtained with agar plate assays should be assessed critically, since such factors as interaction of the test chemical with agar or the nature of diffusion of test materials in the complex colloidal agar medium have a very real effect upon the availability of active material to the fungus. In addition, the assessment of fungal growth requires care, since some fungitoxic materials cause an initial lag period before growth rate becomes linear with time (Bateman, 1933). Trinci (1971) has discussed factors involved in the growth of fungal colonies on solid media.

The measurement of growth in liquid culture requires careful attention to such factors as quantitative separation of fungal material from growth medium and drying under carefully-controlled conditions to achieve a constant weight.

The value of the agar plate test is obvious and it is used as a primary screen by at least one major fungicide testing laboratory in Great Britain. Nevertheless, a thorough elucidation of the fungitoxic properties of a given material *in vitro* requires consideration of data from other investigations which are discussed in the following sections.

Suppression of fungal growth is also assessed in a variety of useful bio-autographic techniques in which fungitoxic compounds separated on paper-chromatograms (Weltzien, 1958; Dekhuizen, 1961) or on thin layer chromatographic plates (Narasimhachari and Ramachandran, 1967; Peterson and Edgington, 1969; Homans and Fuchs, 1970) are visualized by spraying and incubation with fungal spores in a nutrient medium.

Inhibition of spore germination

Discussions of the theoretical and practical aspects of spore germination tests in the laboratory have been presented by Wilcoxon and McCallan (1939) and Horsfall *et al.* (1940). The importance of this technique as derived by McCallan (1930) is emphasized by its being credited with enab-ling the development of a large number of fungicides over the period 1945–1960 (Lukens, 1971) and variations of the technique are still in daily use. Indeed, the slide germination technique for the assessment of inhibi-tion of spore germination was considered to be of such value that a recom-mended procedure was published by the American Phytopathological Society (Anon, 1943). The essence of the method is the application of drops of a spore suspension to glass microscope slides previously coated with a deposit of test chemical, although use of an alternative test-tube dilution technique in which spores are suspended in solutions of the test chemical was also recommended (Anon, 1947). Important factors influencing attain-ment of reproducible results are strict attention to cleanliness of all glass-ware, careful and reproducible deposition of test material upon the slide, and standardization of the fungal inoculum. Thus, the presence of deter-gent residues may affect the pattern and reproducibility of the fungicide deposit whilst the presence of chromium or other metal ions (from cleaning solutions) may drastically alter germination of the spores. Deposition of test material upon the slide is best achieved with a vertical rather than a horizontal spray chamber, such as that described by McCallan and Wil-coxon (1940), whilst less elaborate but equally efficient equipment is used in our laboratory (Hislop, 1968). A valuable discussion on methods of obtaining spore inoculum suitable for germination tests was published by McCallan and Wilcoxon (1939). They emphasized the need for the inclusion of a standard fungicide in each experiment in order to enable valid comparisons of results obtained at different times. Many workers have attempted to standardize the size of the drop of spore suspension applied to the slide, and a useful summary of various papers dealing with this aspect may be found in the review by Neely (1969).

Some obligate parasites such as powdery mildews are not usually employed in spore germination tests because they often germinate poorly in liquid water. Zaracovitis (1964) showed, however, that under near-optimal conditions many powdery mildew conidia will germinate on *clean* glass surfaces and we have regularly used a modification of this technique for laboratory assays with these pathogens (Clifford and Hislop, 1973). It is not always desirable or convenient to germinate spores on glass surfaces, and de Waard (1971*a,b*) has successfully studied the effect of chemicals (particularly systemic fungicides) on the germination of powdery mildew conidia which had been deposited on cellulose acetate films resting on agar.

Modification of respiration

Virtually any metabolic process characteristic of living organisms which can be measured has been considered as a means of assessing fungitoxicity. Respiration is one such process, which although rarely measured in routine assays can be of considerable use in studies of the mechanism of fungitoxicity. However, McCallan, Miller and Weed (1954) noted a wide variation between the degree of inhibition of respiration and of spore germination by a given substance, and Torgeson (1967) pointed out that the concentration of fungicide necessary to inhibit respiration is usually in excess of that required to inhibit growth. It is perhaps worth remembering that at certain concentrations some substances stimulate respiration by uncoupling oxidative phosphorylation.

Evaluation of fungicides in the glasshouse

According to Torgeson (1967), the Boyce Thompson Institute used four main host/pathogen combinations (bean/rust, bean/powdery mildew, tomato/early blight, tomato/late blight) for their primary screening programme. However, fungicide manufacturers in the U.K., because of the decreasing reliance placed on tests *in vitro*, presently tend to use a wider selection of host/pathogen combinations, the more important of which are recorded in Table 1 (Bent, pers. comm.; Evans, pers. comm.). Factors involved in the choice of suitable host/pathogen systems for the assay of foliage (as opposed to soil) fungicides have been well described by Mc-Callan and Wellman (1943*b*), as follows:

TABLE 1. Fungi which have been used in primary and secondary bio-assays in the glasshouse

Type of test	Fungus	Disease	Host
Foliage	*Phytophthora infestans*	Downy mildew—blight	Potato
	Plasmopara viticola	Downy mildew—blight	Vines
	Botrytis fabae	Chocolate spot	Broad beans
	Botrytis cinerea	Grey mould	French beans
	Piricularia oryzae	Rice blast	Rice
	Venturia inaequalis	Scab	Apples
	Sphaerotheca fuliginea	Powdery mildew	Cucurbits
	Podosphaera leucotricha	Powdery mildew	Apples
	Erysiphe graminis	Powdery mildew	Cereals
	Uncinula necator	Powdery mildew	Vines
	Uromyces phaseoli	Rust	French beans
	Puccinia recondita	Rust	Wheat
Seed	*Pythium ultimum*	Damping off	Peas
	Pyrenophora spp.	Leaf stripe/spot	Cereals
	Septoria nodorum	Glume blotch	Wheat
	Fusarium culmorum	Seedling blight	Wheat
	Tilletia caries	Bunt/stinking smut	Wheat
	Ustilago nuda	Loose smut	Barley
Seed/soil	*Fusarium nivale*	Snow mould	Rye
Soil	*Verticillium albo-atrum*	Wilt	Cotton
	Rhizoctonia solani	Damping off/root rot	Cotton
Stored fruit	*Penicillium digitatum*	"Blue mould"	Oranges

"*Host plant.* The host plant should be grown with ease and rapidity, should develop uniformly, be of a fairly open habit of growth suitable for spraying, and robust enough to withstand spray pressure; the surface of the leaves should allow adequate wetting by typical fungicides; the plant should not be unduly sensitive to spray injury; preferably the plant should also represent a crop of importance. Slow-growing plants occupy valuable greenhouse space and are frequently subject to attack by other plant diseases and insect pests."

"*Fungus.* The fungus should be readily cultured and produce an abundance of pathogenic spores in a relatively short time. It should be representative of one of the major groups of plant disease fungi."

"*Disease.* The disease should be readily obtained following simple procedures of inoculation and incubation; it should be reproducible; the lesions should develop rapidly, be seen with ease, and be suitable for quantitative measurement. The disease should not spread under normal greenhouse conditions. Finally, it is preferable that the disease should be of economic importance."

Unfortunately, except in a few cases, practical considerations preclude the meeting of all these conditions. Furthermore, glasshouse screening methods for substances which are toxic to pathogens of fruit trees, for example, must be carried out on seedlings or rootstocks and these may be very different from the trees used in commercial practice. For tests with obligate parasites provision must be made for adequate glasshouse space for incubation of host plants necessary for the production of inoculum.

Having selected an appropriate host/pathogen combination, the basic procedures for testing *in vivo* are the same as those *in vitro*, i.e. application of test chemical, inoculation, incubation and assessment of control. Although test materials and inoculum cannot be applied with the same precision as in the laboratory, it is nevertheless essential to do so as carefully and reproducibly as possible. This is often achieved, both for test chemicals and spore suspensions, by the use of a paint spray-gun mounted in a cabinet fitted with a suitable extractor fan. Spraying is often carried out simultaneously from several positions and optimum cover is facilitated by rotation of the plant on a turn-table (e.g. Hislop and Park, 1962). In commercial testing practice it is usual to apply the compound (at concentrations similar to those employed with currently used commercial fungicides), simultaneously to leaves and soil. Only if the material gives adequate protection of the plant under these conditions is it used in further and different tests. In general, for diseases other than those caused by powdery mildews, it is essential to incubate the inoculated plants under conditions which preclude the drying out of inoculum drops, since liquid water must be present for germination and infection to occur. The latter processes are also enhanced by incubation at a near-optimal temperature for the particular disease involved. Infection usually occurs within 24 h and plants may then be transferred to a suitable glasshouse (preferably equipped with capillary benches, supplementary lighting and heating). Symptoms of infection take varying periods to develop and are best assessed at a fairly early stage.

In the case of diseases caused by powdery mildews, inoculation is conveniently achieved by blowing conidia from infected leaves and allowing them to sediment in a draught-proof settling tower (e.g. Kirby and Frick, 1963).

Useful reviews of methods pertinent to the principles and practices of testing for fungitoxicity against specific host/pathogen combinations have been presented by Hamilton (1959) and Torgeson (1967).

The introduction of a host plant into the bio-assay system allows evaluation of other important factors such as phototoxicity, eradicant activity and movement in the host plant. Formulation, compatibility with other pesticides, the effects of light, humidity, temperature and leaf exudates are also important.

Eradicant activity is conveniently assessed in the glasshouse by applying the test chemical to infected plants, from one to seven days after inoculation, and assessing the degree of control achieved.

Measurement of uptake and translocation of test compounds by roots of the host plant (systemic activity) has become an increasingly important facet of the bio-assay of fungicides, but care must be taken to obviate irreversible adsorption of fungitoxicant by the soil which renders it unavailable to the plant (Hickey, 1969). The choice of host plant for such tests is important, since translocation of fungitoxic materials occurs much less readily in woody than non-woody stems; for example, we have observed facile movement of several commercial systemic fungicides in young cucurbit plants, some movement in apple seedlings, and very limited movement in more mature apple plants. The investigation of uptake into and transmission through leaves of fungitoxic materials is also important. Indeed, translaminar activity, i.e. the movement of active material from the abaxial to the adaxial surface of the leaf or vice versa, may be very important in the control of foliage pathogens which are difficult to control where conventional spraying methods give unsatisfactory cover of foliage with fungicide.

With two possible exceptions, most materials known to be systemic move in the plant's transpiration stream (i.e. upwards and outwards in stems and leaves). However, a substance which moves basipetally in the phloem would be of importance since sprays applied to foliage could be adsorbed into the plant and translocated subsequently into new growth, so that the limitations of translocation in woody stems would be reduced. Hence, it is important to study forward and backward movement of materials in leaves by applying them only to the apical or basal halves and to one single leaf of a growing plant for observations on the protection of new growth (Clifford, 1971).

Some fungitoxic compounds which apparently have no effect on mycelia nevertheless inhibit the formation of spores (Lukens and Horsfall, 1968). This antisporulant activity may be measured in the glasshouse by treating infected plants with the chemical and assessing the degree of sporulation (Byrde and Sztejnberg, pers. comm.).

Measurement of the resistance of protectant fungicides to erosion by artificial rain can be an important facet of glasshouse testing of materials which have survived primary and secondary screens (Burchfield and Goenaga, 1957; Hamilton, 1959; Hislop and Park, 1962). Other aspects of plant protection worthy of study in the glasshouse are redistribution of fungitoxic materials from leaf to leaf by rain (Hamilton, 1959; Hislop, 1969), transference of active material in the vapour phase (Bent, 1967; Hislop, 1967), and improved control resulting from fungicide-leaf surface interactions (Clifford, Hislop and Holgate, 1970).

The treatment of seeds with fungitoxic formulations represents a very important and economical facet of pesticide usage in that infection by seed-borne or soil-inhabiting pathogens may be controlled very successfully at minimal cost; methods for evaluating such materials in the glasshouse are therefore very necessary. In this context two situations are pertinent, viz the use of infected seed sown in sterile soil and of healthy seed sown in infected soil, and the experimental material should be tested both by application to the seed and by incorporation in the soil. These techniques involve considerable difficulties in that application of a uniform dose of material to a seed is rarely achieved and is influenced by factors such as the nature of the seed and of the chemical, methods used to coat the seeds, and formulation of the chemical. Similar problems occur with respect to incorporation of experimental materials into soil, and care is necessary in standardization of the composition and humidity of the particular soil used in tests from day to day. Assessment of the efficacy of these treatments should involve such considerations as number of seedlings which survive, size, vigour and freedom from infection, since these records provide information about phytotoxicity, fungitoxicity and possible plant growth substance activity. Methods for assessment of fungitoxic seed treatments have been published by McCallan (1948), Koehler and Bever (1956), Machacek and Wallace (1957) and Purdy (1958). The last named author extended the duration of his test in order to allow assessment of the number of seed heads which were smutted at maturity.

Soil fumigation using volatile materials is a specialized technique for the control of soil-borne pathogens and has been reviewed by Purdy (1967) and Munnecke (1967). Fumigants are usually applied to soil by injection and the pathogen is affected by the toxicant in the vapour phase (Page, 1963). Many fumigants are general soil sterilants that kill fungi, bacteria and nematodes and it should be realised that they may kill useful as well as harmful organisms. Since many of these materials are phytotoxic, it is often necessary to apply them to soil many days prior to planting. Maximum benefit is obtained by covering the soil surface immediately after treatment.

Testing in the field

That the anatomy and morphology of plants raised in the glasshouse differ from those of plants grown in the field is well known. In addition under glasshouse conditions infection occurs more readily and the control of pathogens is achieved more easily. Since many fungicides will be used in the field, it is obvious that their testing under these conditions is vitally important. Here too, factors such as the weather, unsatisfactory formulation, decomposition of active material by light or by organisms other than pathogenic fungi and potential phytotoxicity may show significantly greater effects than in the glasshouse. Adequate coverage of plant surfaces is achieved easily in the glasshouse but very much less easily in the field, where the type of crop, spraying apparatus, spraying technique and climatic conditions have a marked influence.

Initially, only small quantities of an experimental material are likely to be available since the feasibility of using it as a fungicide must be established before even pilot plant scale production would be contemplated. Thus, early field trials are usually carried out on small plots which are normally established on experimental stations, and involve a statistical design capable of yielding the maximum data for minimum outlay of space, labour and materials. Here the investigator is looking for performance equal to or better than those of all known competitive products, and this is the stage at which the decision is made as to whether or not the material should proceed to larger and more rigorous (and consequently more expensive) field trials.

The final evaluation of a potential fungicide can only be made by correlating data derived from a wide variety of tests including those from larger scale field trials which are usually carried out in three situations, i.e. (1) in carefully regulated statistical trials in the confines of experimental stations; (2) in supervised trials on commercial holdings and (3) in less-sophisticated (but more numerous) trials carried out by selected potential users. Any produce obtained from such trials may not be sold unless official clearance has been received for the use of the material. This usually follows from residue and toxicological studies.

Wide differences in susceptibility to a given pathogen by different cultivars are common, and it is essential that field trials are conducted on a wide range of cultivars of all of the major species of host plants for which the product is likely to be recommended. It is also well known that some cultivars are very much more prone to phytotoxicity than others. Equally, it is necessary to conduct tests over a period of at least three years in geographic locations which differ widely in climate and soil type in order to take maximum account of variability of climatic conditions at a given site.

Uniformity of host plants, consistency of techniques and randomisation and replication are vital factors in the performance of any bio-assay technique, but in field trials they need to be considered in a sophisticated manner if the most useful information is to be derived. In this context, experimental design must take into account factors such as availability and spread of inoculum, variation in soil type within an experimental plot, the topography of the site, variability in size and vigour of plants or seeds and, of course, the physical size of the available plot. On the more practical side, the degree of sophistication of the experiment will be limited by the number of experimental materials to be tested and the nature of the techniques to be used for their application. The amount and skill of the labour available for recording, harvesting and assessment of crop yield and quality are also important factors to be considered. At some stage of this testing programme it may be necessary for the more effective materials to be examined for compatibility with other pesticides. A detailed consideration of experimental design and analysis is beyond the scope of this article, but may be found in the texts of Cochran and Cox (1957) and Snedecor (1956).

Quantification of fungitoxic activity

An unfortunate feature of all biological assay methods is variability in the reaction of the test subjects to a given chemical and the consequent impossibility of reproducing at will the same result in successive trials, however carefully the experimental conditions are controlled. Thus, experimental results may only be interpreted to best advantage if they are subjected to precise and critical statistical examination; indeed, an unsatisfactory treatment may lead to conclusions which are incomplete, unreliable or even actively misleading (Finney, 1947).

Responses of fungi to the presence of toxic materials are either quantitative (e.g. extent of growth of the organism on an agar plate) or of an "all or none" (quantal) nature as in the case of spore germination tests. Laboratory and glasshouse test results may be expressed in one of two ways, i.e. the response resulting from unit dose of toxicant, or the dose of toxicant required to produce unit response. The former has been used extensively (mainly because it involves less labour) but results obtained can be misleading in that a compound which appears to be less active than another at one concentration might be much more efficient at twice this concentration. In field trials (which are necessarily less precise), dosage/response analyses are rarely attempted, and it is more usual to examine results for statistical differences between the performances of experimental materials and presently used fungicides (which are included in the test as standards) and of unsprayed control plants in order to assess the degree of severity of infection.

The percentage germination of fungal spores in a spore germination test for a series of fungicide concentrations plotted against applied concentration will usually yield a skewed curve which may be transformed to a sigmoid one by using the logarithm of applied concentration. However, Gaddum (1933), Bliss (1935) and Bliss and Marks (1939) showed that this sigmoid curve could be converted to a (more useful) straight line by conversion of the percentage response units to units of the standard deviation of the normal distribution (which were called probits).

Various graphic and other rapid statistical methods for determining the significance in differences between treatments have been reviewed by McIntosh (1961). Perhaps the most useful and simple technique is the use of logarithmic probability graph paper where percentage response is plotted directly against dose on a logarithmic scale yielding a dosage-response curve from which an index of fungitoxicity can be derived. The dosages which would be expected to effect a 50% or a 95% response (respectively referred to as the ED_{50} or ED_{95} values) are commonly used in this context. The ED_{50} value is widely used for comparisons of fungitoxic activity *in vitro* since in this case it is considered to be the most precise point of the curve and its determination requires less effort. However, for comparison of activities *in vivo*, the ED_{95} value is preferred since it has a more practical relationship to chemical control of disease in the field and, in this latter situation, is more precisely determined (Horsfall and Barratt, 1945). Sources of non-linearity in dosage-response curves may be attributed to factors such as differing modes of action of the toxicant resulting from varying concentrations of different active species which may be present, variation in manipulative techniques, or differences in the concentration of active species actually reaching the site of action within the organism. It has been suggested that the possession of parallel dosage-response lines by different compounds indicates a common mode of action (Dimond *et al.*, 1941; Horsfall, 1956). McCallan, Burchfield and Miller (1959), however, pointed out that the relationships between uptake and external concentration of various fungitoxic materials were represented by three types of curves and concluded that the slope based on applied concentration did not necessarily indicate mode of action since the mathematical relationship between uptake and applied concentration was not predictable.

In view of the variability shown in biological assays, it is preferable to carry out several tests involving few replicates rather than few tests with many replicates (McCallan and Wellman, 1943a). Furthermore, it is highly desirable to include in each test a standard fungitoxic substance in that the performance of this material will facilitate appraisal of variability of results from test to test.

The principles involved in quantification of results of bio-assays carried out in the glasshouse are essentially the same as those for tests *in vitro*, with one slight difference. Tests *in vitro* involve measurement of some aspect of the development of the fungus in the absence of a host plant whereas in glasshouse tests the severity of infection of the host is measured. Methods of assessment therefore usually involve either counting the number of lesions on, for example, unit area of leaf or estimating the area of leaf which is infected or healthy. Both types of evaluation are facilitated by the use of simple scoring systems based on standard reference keys as reviewed by Moore (1969).

Dosage-response curves may be constructed from data derived from glasshouse tests, and ED_{50} or ED_{95} values calculated. The slopes of the lines are generally less than those for results *in vitro* with the same compounds, and it has been pointed out (McCallan and Wellman, 1943*a*; Wellman and McCallan, 1943) that glasshouse tests are considerably less precise than spore germination tests.

In field assays, dosage-response curves are replaced by more appropriate observations such as degree of protection achieved from the application of a given concentration of active material to a unit area of plots, even though this may have little correlation with the dose of chemical deposited upon a unit area of plant.

The degree of infection of the host plant at various stages of its development is commonly recorded, but the most important factors in field trials are crop yield and quality, since these influence a grower's decision whether or not to use a particular material. An example of this is in cereal crops, where control of powdery mildew infections does not necessarily lead to an economically significant increase in crop yield or quality (Lester, 1971).

Conclusion

The foregoing represents an account of conventional methods for assaying fungicides, but the need for comprehensive study of mechanisms of fungitoxicity has been emphasized over the last few years by the dramatic introduction of systemic fungicides. Thus Schlüter and Weltzien (1971) and Clifford and Hislop (1971) demonstrated that many of these compounds do not inhibit conidial germination but interfere with some later stage in the development of infection. However, Koch (1971) suggested the possibility of predicting systemic fungitoxicity by observing modification of spore germination *in vitro*. We suggest that *in vitro* tests are valuable more for investigation of modes of action of chemicals than for attempted prediction of field performance. Whilst the present state of knowledge permits prediction of chemical structure likely to be associated with fungitoxicity, it is not yet possible to design compounds which are guaranteed to be effective commercial fungicides, and the "synthesise it and test it" procedure will continue to be of importance. Glasshouse testing will continue to be an essential feature of fungicide research.

Acknowledgements

We greatly appreciate the help of Drs E. Evans and K. J. Bent who supplied us with valuable information as to current commercial practice in fungicide testing. The help of Drs R. J. W. Byrde and D. Woodcock in reviewing this paper is much appreciated.

References

ANON. (1943). American Phytopathological Society Committee on Standardization of fungicidal tests. The slide-termination method of evaluating protectant fungicides. *Phytopathology*, **33**, 627.

ANON. (1947). The test tube dilution technique for use with the slide-germination method of evaluating protectant fungicides. *Phytopathology*, **37**, 354.

ANON. (1972). Pesticides in the modern world (pp. 30–35). A Symposium prepared by members of the Co-operative Programme of Agro-Allied Industries with F.A.O. and other Organizations.

BAECHLER, R. H. (1967). Application and use of fungicides in wood preservation, pp. 425–461. In *Fungicides: An Advanced Treatise*. Vol. 1. (D. C. Torgeson, ed.). Academic Press, London.

BATEMAN, E. (1933). The effect of concentration on the toxicity of chemicals to living organisms. *U.S. Dept Agr. Tech. Bull.* No. 346.

BENT, K. J. (1967). Vapour action of fungicides against powdery mildews. *Ann. appl. Biol.*, **60**, 251.

BLISS, C. I. (1935). The calculation of the dosage-mortality curve. *Ann. appl. Biol.*, **22**, 134.

BLISS, C. I. & MARKS, H. P. (1939). The biological assay of insulin. II. The estimation of drug potency from a graded response. *Quart. J. Pharm. Pharmacol.*, **12**, 182.

BLOCK, S. S. (1967). Application and use of fungicides as industrial preservatives, pp. 379–423. In *Fungicides: An Advanced Treatise*. Vol. 1. (D. C. Torgeson, ed.). Academic Press, London.

BURCHFIELD, H. P. & GOENAGA, A. (1957). Equipment for producing simulated rain for measuring the tenacity of spray deposits to foliage. *Contr. Boyce Thompson Inst. Pl. Res.*, **19**, 133.

BYRDE, R. J. W., CLIFFORD, D. R. & WOODCOCK, D. (1958). Fungicidal activity and chemical constitution. VI. The activity of some substituted -8 hydroxyquinolines (oxines). *Ann. appl. Biol.*, **46**, 167.

BYRDE, R. J. W., CLIFFORD, D. R. & WOODCOCK, D. (1961). Fungicidal activity and chemical constitution. IX. The activity of 6-n-alkyl-8-hydroxy-quinolines. *Ann. appl. Biol.*, **49**, 225.

CASARINI, B. & PUCCI, R. (1957). Prove *"in vitro"* della efficacia di diversi anticrittoganici sec funghi scetti fra gliabituali "test" e fra alcuni dannosi parasiti della piante. *La Ricerca Scientifica*, **27**, 1468.

CLIFFORD, D. R. (1971). *The control of powdery mildews by alkyldinitrophenol fungicides*. M.Sc. Thesis, Univ. Bristol.

CLIFFORD, D. R., HISLOP, E. C. & HOLGATE, M. E. (1970). Some factors affecting the activities of dinitrophenol fungicides. *Pestic. Sci.*, **1**, 18.

CLIFFORD, D. R. & HISLOP, E. C. (1971). The activities of some systemic fungicides against powdery mildew fungi. *Proc. 6th Br. Insectic. Fungic. Conf. 1971*, **2**, 438.

CLIFFORD, D. R. & HISLOP, E. C. (1973). Studies *in vivo* and *in vitro* with powdery mildew fungicides. *Ann. appl. Biol.*, **73**, 299.

COCHRAN, W. G. & COX, G. M. (1957). *Experimental Design*, 2nd Ed. John Wiley, London.

CRAMER, H. H. (1967). Plant protection and world crop production. *Pfl Schutz-Nachr. Bayer.* **20**, 524 pp. Farbenfabriken Bayer Ag., Leverkusen, West Germany.

CURTIS, L. C. (1944). The influence of guttation fluid on pesticides. *Phytopathology*, **34**, 196.

DEKHUIZEN, H. M. (1961). A paper chromatographic method for the demonstration of fungitoxic transformation products of sodium dimethyldithiocarbamate in plants. *Meded. Landbhogeschl OpzoekStns, Gent*, **26**, 1542.

DIMOND, A. E., HORSFALL, J. G., HEUBERGER, J. W. & STODDARD, E. M. (1941). Role of the dosage-response curve in the evaluation of fungicides. *Conn. Agr. Expt Stn, New Haven. Bull. No.* **451**.

FINNEY, D. J. (1947). *Probit Analysis—a Statistical Treatment of the Sigmoid Response Curve*. Cambridge University Press, Cambridge.

FORSBERG, J. L. (1949). A new method of evaluating fungicides. *Phytopathology*, **39**, 172.

GADDUM, J. H. (1933). Report on biological standards. III. Methods of biological assay depending upon a quantal response. *Spec. Rept Ser. Med. Res. Coun. Lond.*, No. **183**.

HAMILTON, J. M. (1959). Evaluation of fungicides in the greenhouse, pp. 253–257. In *Plant Pathology—Problems and Progress, 1908–1958*. University of Wisconsin Press, Madison.

HICKEY, K. D. (1969). Apple powdery mildew (*Podosphaera leucotricha*). *Fungicide-Nematicide Tests*, **25**, 20.

HISLOP, E. C. (1967). Observations on the vapour-phase activity of some foliage fungicides. *Ann. appl. Biol.*, **60**, 265.

HISLOP, E. C. (1968). *Redistribution of fungicides on plants*. Ph.D. Thesis, University of Bristol.

HISLOP, E. C. (1969). The effect of rain on plants, pests and pesticides. The redistribution of fungicides by rain. *Chemy Ind.*, **42**, 1498.

HISLOP, E. C. & PARK, P. O. (1962). Studies on the chemical control of *Phytophthora palmivora* Butl. on *Theobroma cacao* L. in Nigeria. II. Persistence of fungicides on pods. *Ann. appl. Biol.*, **50**, 67.

HOMANS, A. L. & FUCHS, A. (1970). Direct bioautography on thin-layer chromatograms as a method for detecting fungitoxic substances. *J. Chromatog.*, **51**, 327.

HORSFALL, J. G. (1945). Quantitative bioassay of fungicides in the laboratory. *Bot. Rev.*, **11**, 357.

HORSFALL, J. G. (1956). *Principles of Fungicidal Action.* Chronica Botanica Co., Waltham, Mass.

HORSFALL, J. G. & BARRATT, R. W. (1945). An improved grading system for measuring plant diseases. *Phytopathology*, **35**, 655.

HORSFALL, J. G., HEUBERGER, J. W., SHARVELLE, E. G. & HAMILTON, J. M. (1940). A design for the laboratory assay of fungicides. *Phytopathology*, **30**, 545.

HORSFALL, J. G. & RICH, S. (1951). Fungitoxicity of heterocyclic nitrogen compounds. *Contr. Boyce Thompson Inst. Pl. Res.*, **16**, 313.

KENDRICK, J. B. & MIDDLETON, J. T. (1954). The efficacy of certain chemicals as fungicides for a variety of fruit, root and vascular pathogens. *Pl. Dis. Reptr.*, **38**, 350.

KIRBY, A. H. M. & FRICK, E. L. (1963). Greenhouse evaluation of chemicals for the control of powdery mildews. I. A method suitable for apple and barley. *Ann, appl. Biol.*, **51**, 51.

KOCH, W. (1971). Behaviour of commercial systemic fungicides in conventional (non-systemic) tests. *Pestic. Sci.*, **2**, 207.

KOEHLER, B. & BEVER, W. M. (1956). Effect of fungicide and storage temperature on fungicide injury to wheat seed. *Pl. Dis. Reptr,* **40**, 490.

KUHFUSS, K. H. (1957). Beitrag zur methodik der Fungizidprufung von Nass- und Frockenbeitzmitteln. *Phytopath. Z.*, **28**, 281.

LEBEN, C. & KEITT, G. W. (1950). A bioassay for tetramethylthiuramdisulphide. *Phytopathology*, **40**, 950.

LESTER, E. (1971). Cereal leaf disease control—practicable and economic considerations. *Proc. 6th Br. Insectic. Fungic. Conf. 1971*, **3**, pp. 643–647.

LUKENS, R. J. (1971). *Chemistry of Fungicidal Action.* Springer-Verlag, Berlin.

LUKENS, R. J. & HORSFALL, J. G. (1968). Glycolate oxidase, a target for antisporulants. *Phytopathology*, **58**, 1671.

MACHACEK, J. E. & WALLACE, H. A. H. (1957). A study of the germination of barley seed treated to control loose smut. *Can. J. Pl. Sci.*, **37**, 59.

MANTEN, A., KLÖPPING, H. L. & VAN DER KERK, G. J. M. (1950). Investigations on organic fungicides. II. A new method for investigating antifungal substances in the laboratory. *Antonie van Leeuwenhoek J. Microbiol. Serol.*, **16**, 282.

MARSH, R. W. (1936). Notes on a technique for the laboratory evaluation of protective fungicides. *Trans. Br. mycol. Soc.*, **20**, 304.

MASON, C. L. & POWELL, D. (1947). A *Pythium* plate method of evaluating fungicides. *Phytopathology*, **37**, 527.

McCALLAN, S. E. A. (1930). Studies on fungicides. II. Testing protective fungicides in the laboratory. *Cornell Univ. Agr. Expt Sta. Mem. No. 128.*

McCALLAN, S. E. A. (1947). Bioassay of agricultural fungicides. *Boyce Thompson Inst. Pl. Res., Professional Papers.*, **2**, 23.

McCALLAN, S. E. A. (1948). Evaluation of chemicals as seed protectants by greenhouse tests with peas and other seeds. *Contr. Boyce Thompson Inst. Pl. Res.*, **15**, 91.

McCALLAN, S. E. A. (1959). Evaluation of fungicides in the laboratory, pp. 248–253. In *Plant Pathology—Problems and Progress 1908–1958.* University of Wisconsin Press, Madison.

WELTZIEN, H. C. (1958). Ein biologisches test fur fungizide substanzen auf dem Papierchromatogramm. *Naturwiss*, **45**, 288.

WILCOXON, F. & MCCALLAN, S. E. A. (1939). Theoretical principles underlying laboratory toxicity tests of fungicides. *Contr. Boyce Thompson Inst. Pl. Res.*, **10**, 329.

YARWOOD, C. E. (1945). Copper sulphate as an eradicant spray for powdery mildew. *Phytopathology*, **33**, 1146.

ZARACOVITIS, C. (1964). Factors in testing fungicides against powdery mildews. The germination of conidia *in vitro*. *Annals. Inst. Phytopath. Benaki*, **23**, 73.

ZENTMEYER, G. A. (1955). A laboratory method for testing soil fungicides with *Phytophthora cinnamomi* as test organism. *Phytopathology*, **45**, 389.

Testing Biocidal Paints

W. ROBERT SPRINGLE

*Paint Research Association, Waldegrave Road, Teddington,
Middlesex TW11 8LD, England*

Paint is frequently the last material applied directly to many surfaces. It may, therefore, often be conveniently used as a semi-permanent means of controlling biological growth which might otherwise develop on the surface causing disfigurement and, sometimes, damage. Although perhaps more generally obvious in tropical regions, mould growth on painted and coated surfaces of buildings (both internal and external) is an ubiquitous problem. This growth is related to many, often specific, factors the most important of which is water in the form of condensation or dampness. Besides disfiguring, fungal growth can cause damage to materials and either be, or indicate, a hazard to health.

Since algae require light as well as inorganic salts and abundant water for growth, they occur normally on exterior surfaces of buildings and structures, particularly those constructed of concrete or rendered with cement, mortar or lime-based materials. Algal disfigurement frequently occurs along water runs and its distribution often shows a marked dependence on constructional and architectural factors.

Many problems of fungal or algal growth could be dealt with by improving ventilation, introducing structural changes, or removing sources of nutriment. An alternative method of effecting protection is to incorporate a biocide into the coating formulation. The Paint Research Association has therefore found it necessary to evolve methods of evaluating paints and coatings for fungicidal and algicidal effectiveness.

General techniques

A technique such as that described below is not easily quantified; its standardization is currently the subject of an International Committee on Biodeterioration. In using it to assess biocidal activity of paint films and coatings, conditions are selected according to the following:

Nature of substrate which carries the test coating

In assessing biocidal activity at the Paint Research Association (PRA), a glass boiling tube (*c.* 15×2·5 cm), either with an abraded surface to improve adhesion, or overlaid with materials such as plaster, mortar, cement, lining paper, is used. The choice of substrate depends on the type of coating being tested, as does the method of application (e.g. paints could be applied with a brush, as a spray or dip).

Pre-conditioning of coating before biological exposure

Once a film is dry it may be tested without any further pre-treatment (i.e. unweathered), or it may first be aged by leaching, heating or artificial weathering. Again the choice depends on the type of coating and its probable situation in use.

Nature of inoculum

Ideally, coatings should be inoculated with an organism, or mixture of organisms which they are likely to meet in use. Alternatively, they may be infected with a standard organism or mixture of organisms. The PRA normally follows this latter approach, basing choice of organisms on information obtained by examining numerous examples of infected coatings.

Experimental conditions in test cabinet

After inoculation, coated-tubes are incubated in specially designed cabinets which allow control of temperature, humidity, illumination and atmospheric composition. Unlike tropical chamber methods, these cabinets give controlled condensation on the inoculated surface (by maintaining an inside/outside temperature differential) thus producing conditions conducive to fungal or algal growth on susceptible materials.

The Fungal test cabinet

This simple, robust apparatus (Fig. 1) consists (Skinner, 1970, 1972*a*) of a stainless steel base tank (30×22×6 cm) sealed with a plastic gasket to a partitioned Perspex top (30×22×15 cm). Water in the base tank is heated by a 120 W heater controlled by an energy regulator on the front panel of the instrument. This allows temperature and humidity inside the cabinet to be maintained at levels conducive to fungal growth. Ten glass tubes used

to carry test paints are suspended in the cabinet through apertures cut in the Perspex top. Controlled condensation is produced on these painted tubes by setting the interior cabinet temperature 2–5° higher than that of surrounding air (the PRA cabinets operate in a room maintained at 21°).

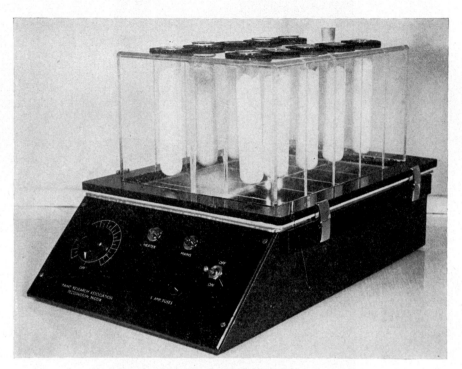

FIG. 1. The fungal test cabinet.

Exterior paints

As an illustration, the procedure used to assess an alkyd gloss paint (a solvent based, drying oil type) for fungicidal activity is described. Rimmed glass test tubes ($15 \times 2 \cdot 5$ cm) are grit blasted or abraded to produce a roughened surface which, after cleaning, considerably assists adhesion. The test paint is coated on to tubes of this type by brushing, spraying or dipping as appropriate and allowed to dry for seven days under controlled conditions. It is then subjected to a period of artificial weathering (e.g. 250 h, rotating daily through 180°) involving water spray and ultra-violet irradiation according to BS 3900, part F3.

After ageing the test paint is dusted with sterile soil particles, sprayed

FIG. 2. A photograph of a susceptible gloss paint tube. ×1·5

with a mixed fungal spore suspension and incubated in the fungal test cabinet for 21–28 d. Spore suspensions are prepared in sterile 0·01% (v/v) Tween 80 solution from 14 d old slope cultures on Czapek-Dox agar (Oxoid), and in the case of exterior paints, the inoculum consists of: *Alternaria alternata* (*A. tenuis*), *Aureobasidium pullulans, and Cladosporium herbarum.*

This inoculum should contain at least 10^4 spores/ml and may conveniently be sprayed from an aerosol (e.g. a nose and throat spray). Spraying should be carried out with the test paint in a confined space such as a

closed box and it has been found advisable to filter suspensions through sterile muslin to remove larger lumps or aggregates. Also it is thought that soil particles on the paint film act as a physical entrapment for spores, helping to retain them on smooth surfaces; in practice soil and vegetable debris normally accumulate on exterior coatings.

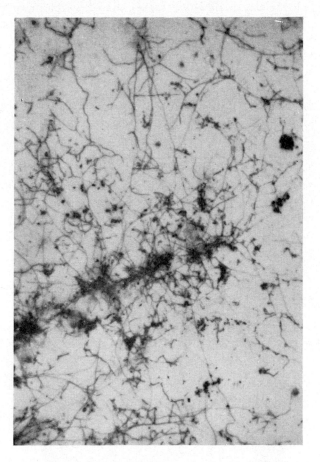

FIG. 3. Mycelial growth on gloss paint. ×25

In each cabinet known control paints accompany test paints, and these are examined regularly for the presence of fungal growth which will occur abundantly on a non-preserved paint in 3–4 weeks. Final assessment is carried out using a stereoscopic microscope. Examples of a susceptible paint and of the prostrate mycelial growth normally obtained on gloss paints are shown in Figs 2 and 3.

Interior paints

By way of illustration, the procedure used to test a water-based emulsion paint for fungicidal activity is described. Rimmed glass test tubes ($15 \times 2 \cdot 5$ cm) are abraded, cleaned and then further coated with a paste of gypsum plaster (e.g. "Sirapite") in 50:50 emulsion binder: distilled water. When dry the new substrate then becomes a layer of plaster c. 1 mm thick which should not fall off when kept damp for an extended period. As in the

FIG. 4. A photograph of a susceptible emulsion paint tube. $\times 1 \cdot 5$

general technique described earlier, the test paint is applied by an appropriate method and allowed to dry for 48 h under controlled conditions.

The dry paint is then tested, either without any pre-ageing or after leaching for a set time with distilled water, by inoculating with a mixed fungal spore suspension and incubating in the test cabinet for 21–28 d. In this case spore suspensions, prepared as before, consist of: *Cladosporium herbarum, Paecilomyces variotii,* and *Stemphylium dendriticum.*

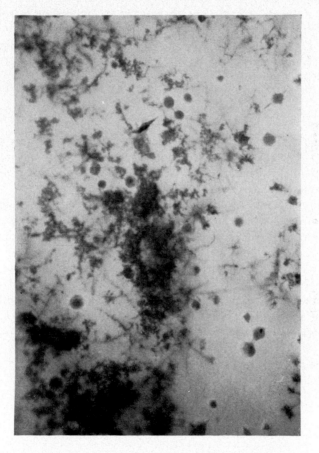

FIG. 5. Colonial growth on emulsion paint. ×25

The technique used is similar to that already described except that for interior paints the surface is not dusted with soil particles. Examples of a susceptible paint and of the erect colonial growth often obtained on emulsion paints are shown in Figs 4 and 5.

Algal test cabinet

The algal test cabinet (Fig. 6) is essentially similar in principle and general design (Skinner, 1972*b*) to the fungal test cabinet (Fig. 1). In this case, however, the Perspex top is unpartitioned and has a gas inlet at the rear which allows gas mixtures, other than normal air, to be introduced. Also the cabinet fits into a base which provides two pairs of 8 W daylight fluorescent tubes as a source of light energy for photosynthesis. Ten glass tubes used to carry test paints are suspended in the cabinet from apertures cut in the Perspex top. Condensation is produced on these painted tubes by setting the interior temperature 4–6° higher than that of the surrounding air (the PRA cabinets operate in a room maintained at 21°).

Exterior paints

Terrestrial algal growth normally occurs on painted surfaces which are exposed to daylight and are frequently or permanently wetted. As an illustration, the procedure used to assess exterior paints generally for algicidal activity is described.

Rimmed glass test tubes (15×2.5 cm) are abraded, cleaned and further coated with a paste of mortar (i.e. 2 of sand to 1 of cement) in 50:50 emulsion binder:distilled water. The new substrate thus becomes a layer of mortar *c*. 1 mm thick which is leached or artificially weathered for a period to reduce alkalinity (this is of course open to variation). The test paint is applied to this substrate by an appropriate method and allowed to dry for a suitable time under controlled conditions. It is then subjected to a period of artificial weathering (e.g. 250 h, rotating daily through 180°) involving water spray and ultraviolet irradiation according to BS 3900, part F3.

After ageing the test paint is dusted with sterile soil particles, sprayed with an algal cell suspension and incubated in the algal test cabinet for

TABLE 1. Media used to grow algae

Medium A		Medium B	
KNO_3	1·21 g	Soil extract	100 ml
$MgSO_4.7H_2O$	2·46 g	K_2HPO_4	0·02 g
K_2HPO_4	1·23 g	$MgSO_4.7H_2O$	0·02 g
$Fe_2(SO_4)_3$	0·05 g	Agar	10·0 g
Sodium citrate	0·19 g	Distilled water	1000 ml
Agar	10·0 g		
Distilled water	1000 ml		

FIG. 6. Algal test cabinet.

28–42 d. Cell suspensions are prepared in sterile distilled water from cultures grown on media such as media A and B in Table 1.

The inoculum normally used by PRA is either *Pleurococcus* sp. (a green unicell) or a mixture of *Pleurococcus* sp. and *Oscillatoria* sp. (a blue-green filamentous). An atmosphere of 5% carbon dioxide in air inside the cabinet appears to promote algal growth in its early stages and a cylinder containing this mixture is attached to the gas inlet.

In each cabinet known control paints accompany test paints, and these are examined weekly for the presence of algal growth. Final assessment is carried out using a stereoscopic microscope (although this is often unnecessary where growth is heavy and bright green). Examples of paints tested by this method are shown in Fig. 7, in which the tube marked 'C' is included only as a comparison.

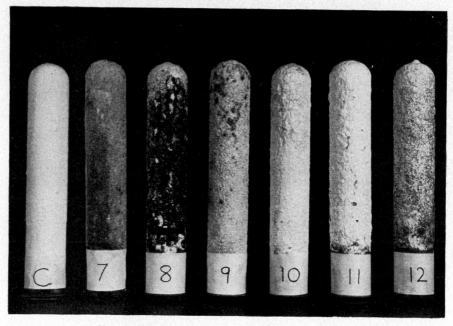

FIG. 7. Emulsion paint tubes from algal test cabinet.

Other coatings

It will be apparent that the principle of the humidity cabinet technique is not limited only to studying paints. It has been used to examine growth resistant properties of cement and mortar, plastic film and paper wall-coverings. Similarly, by choice of suitable pre-ageing conditions the effects of different environments and climatic conditions on test coatings can be simulated and performance against a range of organisms can be determined if the inoculating species are varied.

Assessment

Once cabinet incubation has been completed it is necessary to assess the amount of fungal or algal growth as a measure of the fungicidal or algicidal performances of the test coatings under examination. At present PRA assign a single numerical rating to growth, determined by inspecting the paint surface with a stereoscopic microscope. Often, photographs of the whole tube and "representative" areas on it magnified $\times 25$ are also taken for record purposes, (e.g. Figs 2 and 3). Rating is carried out on a 0–5 scale where:

$0 =$ no growth on test coating
$1 =$ trace of growth on test coating
$2 = 1$–10% coverage of surface
$3 = 10$–30% coverage of surface
$4 = 30$–70% coverage of surface
$5 = $ Over 70% coverage of surface

There is some difference of opinion as to whether it is meaningful to divide the scale from 3–5 since any substantial amount of growth suggests that the test coating is susceptible.

A single, visually-determined numerical rating can really only indicate the amount of growth—telling nothing of its morphology, extent of development and pigmentation. This last point is particularly important to the paint industry since aesthetic considerations are often of fundamental significance. At present it is normal practice to comment on appearance in reporting results, but eventually either the scale may be modified, or an alternative method of assessing growth may be devised.

Acknowledgements

The author wishes to thank the Director and Council of the Paint Research Association for permission to publish this paper.

References*

HENDEY, N. I. (1962). *JOCCA*, **45**, 5, 343.
O'NEILL, L. A. & SKINNER C. E. (1966). *SCI Monograph* **23**, 170–8.
O'NEILL L. A. & SKINNER C. E. (1968). *FATIPAC Congress, Sect. 2, 12–14.*
SKINNER C. E. (1970) *Paint, Oil and Colour J.*, **157**, 177–80.
SKINNER C. E. (1972) *Biodeterioration of Materials*, Vol. 2, Part VIII, pp. 346–54.
SKINNER C. E. (1972) *FATIPEC Congress*, 421–27.
* These references are intended as an introduction to the literature.

Microbiological Assay of Chemicals for the Protection of Wood

A. F. Bravery

Building Research Establishment, Princes Risborough Laboratory, Princes Risborough, Aylesbury, Bucks, England

Introduction

A useful general account of timber decay and its prevention which includes the principles of laboratory testing of wood preservatives against fungi is given by Cartwright and Findlay (1958). Three main factors influence the performance of wood preserving chemicals: (a) the preservative itself, (b) the method of treatment, and (c) the environment in which the treated wood is to be exposed (Smith, 1965; Gibson, 1966). Complete evaluation of chemicals for the protection of wood thus requires data from a number of different tests, some of which can be conducted in the laboratory and others which cannot.

The laboratory testing of wood preservative effectiveness presents several special difficulties. Firstly, chemicals and treatments of potential practical value must normally protect the substrate for several decades and laboratory tests must therefore greatly accelerate the deterioration processes involved. Secondly, wood is a highly complex and variable material and timber species vary considerably in their chemical and physical properties. Thirdly, treated timber will be exposed in different service environments each with its own degree of hazard from various biological organisms or populations of organisms, possibly having different susceptibilities to preservatives. It is impracticable to simulate the full range of environmental conditions and natural populations of micro-organisms in the laboratory, so for certain tests a limited number of individual organisms must be selected for laboratory use. Fourthly, the performance of preservatives and treatments in timbers of commercial sizes cannot be tested readily in the laboratory so that sampling becomes necessary.

Microbiological laboratory assay tests can permit comparative assessments of initial toxicity of the chemicals towards wood-attacking fungi and residual toxicity after exposure to leaching and evaporation. However it is not possible from these tests alone to estimate the precise value of any chemical in practice nor to forecast the probable life of timber treated in any particular way. Furthermore, the amounts of chemical required to inhibit fungal attack in laboratory tests cannot yet be correlated reliably with those required in a particular service environment.

The range of tests employed currently in assaying chemicals for the protection of wood are categorized in the accompanying chart (Fig. 1). Agar tests are used for rapid initial screening, wood block tests are more complicated and vary according to the specific physiological requirements of the different wood-attacking fungi. Finally, at the bottom of the chart the role of development work is indicated as improving the speed, simplicity and reliability of the methods.

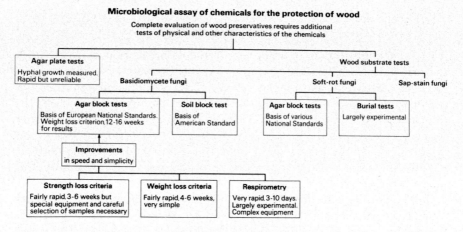

FIG. 1. Chart showing range of laboratory tests for microbiological assay of chemicals for the protection of wood.

Agar Plate Tests

The earliest attempts to standardize a laboratory test for determining the fungal toxicity of wood preservatives (Humphrey and Fleming, 1915; Schmitz *et al.*, 1930), were based on measuring suppression of hyphal growth on nutrient agar containing dilutions of the test chemical (Fig. 2). However, at an early stage the reliability of these agar methods was questioned and comparative studies between agar and wood block tests revealed considerable discrepancies in toxic levels for certain chemicals (Rabanus, 1931; Flerov and Popov, 1933). These discrepancies, and those reported by later authors (Finholt, Weeks and Hathaway, 1952; Smith and Savory, 1965; Unligil, 1972) have been attributed to differences in the extracellular enzyme secretions of the test fungi under different test conditions and to the absence in agar tests of preservative/wood substrate interactions.

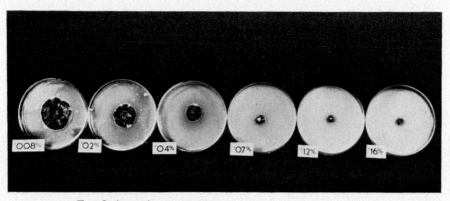

FIG. 2. Agar plate screening test (note toxic point is *c.* 0·1 %).

Early attempts to overcome these inadequacies by using sawdust in agar (Rhodes and Gardner, 1930) were unsuccessful because of difficulties in estimating the amounts of growth. Problems also arise in obtaining homogeneous dispersion of oily chemicals in sawdust/agar substrates. Rabanus (1931) drew attention to another source of variation resulting from altering the proportion of inoculum to poisoned substrate; apparently the presence of an excess of nutrient material in the untreated inoculum has an important influence on the results. Current development work at the Princes Risborough Laboratory (PRL) aims to determine the precise influence of these major sources of variability and to develop an acceptable compromise test which can, if possible, be correlated with wood block tests.

Wood Substrate Tests

In this class of tests, the preserving chemical and solid wood are brought into close contact, to facilitate interactions that may have toxicological significance and ensure that the fungus is in an appropriate physiological state.

There are 4 main groups of lignicolous fungi:
(a) Surface moulds which discolour moist wood surfaces.
(b) Sap-stain (or blue-stain) microfungi, having pigmented hyphae which discolour sapwood.
(c) Wood-destroying microfungi which attack cellulose.

(d) Wood-destroying Basidiomycete fungi capable of attacking wood cellulose and sometimes lignin too.

Surface Moulds

Chemical treatments are not normally employed solely to control mould on timber. The standard procedure for determining the natural resistance to mould growth of manufactured building materials (BS: 1982, 1968) could, in principle, be employed should tests of chemical treatments be required. In this method, test samples are sprayed with a mixture of spores from 5 mould species and incubated in a constant humidity chamber. A new procedure for testing chipboard has been developed at Princes Risborough Laboratory in which test pieces are incubated in contact with moist sawdust (Fig. 3). All assessments are subjectively made by visual observation of overgrowth. Currently, various attempts have been made to quantify results and minimize the subjectivity (Grant, 1973).

FIG. 3. A mould resistance test for chipboard ($\times 0\cdot3$).

Sap-stain fungi

Sap-stain fungi are of economic importance in felled logs, sawn timber and on worked wood in service.

The basic principles of laboratory tests for assessing efficacy of anti-stain chemicals are applicable to each situation although most work has been concerned specifically with log and lumber protection. Hatfield *et al.*

FIG. 4. Unseasoned wood disc test for anti-blue-stain chemicals ($\times 0.5$).

(1950) proposed the unseasoned wood disc method in which 70–80 mm diameter discs of fast-grown sapling logs are dipped in the treating solutions, sprayed with a suitable spore suspension, stacked alternately with untreated controls and incubated in large, closed boiling tubes (Fig. 4). Schulz (1951) used 15 mm cubes of green pine sapwood impregnated under vacuum with the required chemical solution, inoculated with spore suspension and incubated on filter papers in conical flasks. Cserjesi and Roff

(1970) developed a test more specifically applicable to lumber protection in which small boards $(70 \times 17 \times 7$ mm) were cut from green sapwood, dipped, inoculated and incubated over damp filter paper in Petri dishes.

Schulz (1952) evaluated the fungicidal effectiveness of primers intended for use under paint or varnish films using boards $(100 \times 50 \times 9$ mm) of planed softwood artificially infected with staining fungi. The 100×50 mm surface was marked off transversely into 3 equal areas and the test preparation applied to the centre zone. The anti-stain effectiveness of the treatment was compared with the two untreated end zones after incubation. Butin (1961) improved the method by infection with blue-stain spores after preservative treatment and by using a board equally divided by a transverse 3 mm groove to yield two 45×45 mm test surfaces one of which was treated, the other not. The Butin method, with certain minor modification, is likely to form the basis of a new European Standard test for anti-stain wood preservative treatments intended for use both with and without subsequent painting or varnish.

Dickinson (1971) developed a simpler and quicker screening test in which pairs of treated sapwood veneers $(100 \times 25$ mm) were bonded on to an aluminium centre laminate, artificially weathered, sprayed with a mixed spore suspension and then incubated in a humid growth cabinet adapted from that of O'Neil and Skinner (1968).

A visual assessment of chemical effectiveness is employed in all tests of anti-stain treatments, discolouration of treated samples being accepted as evidence of treatment failure.

Soft-rot fungi

Soft-rot of timber is caused by cellulose-destroying microfungi belonging to the *Asomycetes* and *Fungi Imperfecti* (Savory, 1954 *a, b*). They are common soil organisms and assume particular importance in soil contact or in water when the growth of Basidiomycetes is inhibited by moisture contents which are too high or too variable. They can in general tolerate higher concentrations of toxic chemicals than can Basidiomycetes. Although individual fungi can be isolated and shown in pure culture to cause typical soft-rot, under natural conditions soft-rot decay results from the action of a complex of different organisms probably including bacteria. Selection of suitable organisms for laboratory work is complicated because: (a) ecological studies have not identified consistently any particular causal species, and (b) few of those isolated are sufficiently active in pure culture for laboratory test purposes, an issue of particular difficulty with softwood timbers. *Chaetomium globosum* has become widely accepted as a convenient laboratory test organism for hardwoods, though its precise practical relevance is uncertain.

Because of these difficulties, two classes of soft-rot test have emerged. One uses pure cultures of an organism on nutrient agar (Fig. 5) and forms the basis of the British Standard Method (BS: 838, 1961). The second class involves the burial of test blocks in soil or vermiculite substrates. In the case of soil, the natural microflora may sometimes be relied upon to effect decay, e.g., Nordic Wood Preservation Council Standard (NWPC 1.4.1.2/70) otherwise sterile substrates are inoculated either with pure cultures, mixed cultures or with unknown natural microfloras from unsterile soil. The methods of determining the effectiveness of wood preservatives against soft-rot fungi have been reviewed by Savory and Bravery (1970*b*).

FIG. 5. British Standard test against soft rot ($\times 0.7$).

Pure-culture agar tests have the advantage of being better defined than soil tests but it is not yet possible to evaluate chemicals for protection of softwoods by such methods. Whilst this is possible in natural soil burial tests, definition of soils then poses problems. Most test procedures use thin wood samples since mineral salts have to be supplied by direct contact with the substrates in order to provide the additional nitrogen required by soft-rot organisms.

Chemical effectiveness is normally assessed by estimating fungal attack, measured as the loss in weight of test blocks after the required incubation period, and is expressed as toxic values (tv) (formerly known as toxic limits). The tv give the range between the concentration of chemical which just inhibits decay and that which just permits it. A small amount of weight loss is always caused by factors other than fungal attack and this is allowed for in determining the tvs. In Britain and the USA 3% is allowed for experimental error whilst in Germany 5% may be used.

Measurements of loss in strength have also been used in attempts to develop a more sensitive test (Zycha, 1964; Kirk and Schulz-Dewitz, 1968; Wälchli, 1969; Smith, 1970). Toxic values determined using weight loss and strength loss criteria are not always comparable because of other inevitable changes in test procedures. In general the weight loss criterion, though less direct and less informative for research purposes, is more

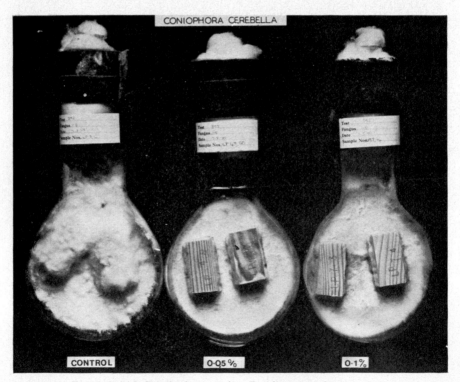

FIG. 6. British Standard test against Basidiomycete fungi (×0·4).

convenient and is thus likely to be preferred for international standardisation of routine tests.

Basidiomycete fungi

The traditional agar/wood block test for determining the fungal toxicity of wood-preserving chemicals was developed early in the 1930s (Liese *et al.*, 1935) and forms the basis of both German and British Standard tests

FIG. 7. American Standard soil block method (×0·6).

(DIN 52.176, 1961; BS 838: 1961) and a draft European Standard currently under discussion by Working Group 38 of the Comité Européen de Co-ordination des Normes. Alliot (1945) developed a similar method employing much smaller wood samples, $30 \times 10 \times 5$ mm instead of $50 \times 25 \times 15$ mm, exposed in test tubes instead of large glass Kollé flasks (the basis of the French Standard tests NFX 41.512, 1961 and NFX 41.502, 1961).

In these tests replicate wood blocks are treated with a series of dilutions of preservative solution, dried, sterilized and exposed to attack by pure cultures of active wood-destroying Basidiomycetes growing on Malt Agar (Fig. 6). Weight losses occurring during incubation are measured and toxic values determined. The tests permit determination of initial toxicity and residual toxicity after leaching procedures but the results are difficult to interpret in practical and commercial terms.

A modified wood block test was developed in America by Leutritz (1946). Impregnated wood samples were exposed to fungal cultures growing on thin strips of wood resting on damp soil in screw top glass jars (Fig. 7). This soil/block method now forms the basis of the American Standard procedure (ASTM, D1413–61, 1961). The soil/block method has advantages over the agar/block method in that a more even moisture content is maintained, volatile toxic constituents of preservatives are absorbed and the wood feeder strip is a more natural source of infection. It is also often claimed that toxic values more closely approach those found in the field (Duncan, 1953, 1958). However, soil is more time consuming in its preparation and variations in its source may influence test results. In general, it would appear that reliable and comparable toxic values can be obtained by either method.

The Nordic Standard soil block test (NWPC Standard 1.4.1.1/70) uses 20 mm cubes treated with dilutions of preservative and embedded in sterilized garden soil in glass jars. After a further sterilization, the jars are inoculated with pure cultures of the test fungi and weight losses incurred over a 12-week incubation period are determined. Two preservative treated blocks and two blocks without preservative are exposed within each jar so that an untreated woody substrate is available. However, toxicity results would be likely to vary somewhat when samples of different soils, or soils from different sources were employed, and more importantly preservative losses are likely to occur during steam sterilization.

Developments

A wide range of different test methods has been used in various studies of wood preservative toxicity, wood decay capabilities of fungi and decay resistance of wood and wood products. The methods differ according to the principal objectives of the researcher. The major test variables are size of wood sample, fungal species or source of infection, nutrient substrate, incubation period and criterion of attack, as cited in many reviews (Colley, 1953; Cartwright and Findlay, 1958; Hartley, 1958; Bravery, 1968; Levi, 1969; Savory and Bravery, 1970a; Bravery and Grant, 1971; Bravery and Lavers, 1971; Hilditch and Hambling, 1971).

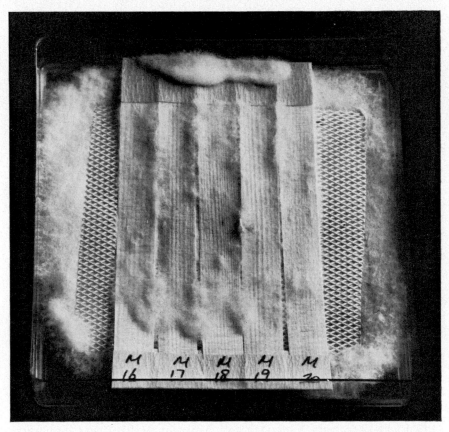

FIG. 8. Petri dish exposure system for this strip tensile test (×1).

The major objectives in developmental research on wood preservative test methodology have been: (a) to reduce the time required to obtain results, and (b) to improve the simplicity and reliability of the methods. Another important objective, more precise interpretation of laboratory results, depends upon a better understanding of the individual and complex combined effects of the different test variables. Work in this direction by Smith (1970) using a soil contact system and a tensile strength criterion has already shown that the period of protection is governed by the preservative concentration, with sample thickness and temperature also having important effects. A significant observation has been that once initiated, the rate of decay in treated wood is fairly rapid and only slightly reduced from that in untreated timber except at very high concentrations of preservative.

The nationally standardized tests use comparatively large wood samples and therefore require exposure periods of 3–4 months in order to develop sufficiently high levels of decay. Although a somewhat smaller test block is used in the French Standard tests, 16 weeks' exposure is still required. Most standard tests also require large, expensive and cumbersome culture vessels and incubation facilities.

The weight loss criterion for assessing timber decay has two important advantages. Firstly, laboratory apparatus and procedures are simple. Secondly, providing blocks are carefully machined to constant volume (Savory and Bravery, 1970), sample preparation and selection need be less stringent than for strength criteria. Findlay (1953) and Smith (1970) have shown that reduction of wood block size increases the rate of decay in a given time and for the purpose of a small scale toxicity test, reduction in block size should permit assessments to be made after only a few weeks' exposure. Rather more careful sample selection than in conventional standardized weight-loss tests is required to exclude fast-grown material with few annual rings. Wood block toxicity tests have recently been developed using small veneers (Levi, 1969) and centimetre cubes (Bravery, 1970) exposed to fungi in petri dish culture (Figs. 9 and 10). These methods have been shown to give results comparable with those obtained in the British Standard test for certain preservatives (Baker, Morgan and Savory, 1972) though this comparability may not prove to be as consistent for all types of preservatives.

All of the standardized methods utilize weight loss as the criterion of decay but significant losses in certain strength properties can be developed during decay more rapidly than losses in weight (Hartley, 1958). Scheffer (1936) and Cartwright, Campbell and Armstrong (1936) have monitored progressive decay by Basidiomycetes and found impact strength to be the most sensitive to decay. Armstrong and Savory (1959) subsequently used the same criterion for assessing decay by the soft-rot fungus *Chaetomium globosum*. Mateus (1954, 1957) evolved a system for measuring changes in the deflection during decay of very much smaller test beams. His system is simple and convenient in that it utilizes Petri dishes (Fig. 11), and his apparatus for measuring deflection has been shown to give a good appproxi-

FIG. 9. Petri dish/wood block test for Basidiomycete fungi (×0·7).

FIG. 10. Petri dish/wood block test for sof-trot fungi (×0·7).

FIG. 11. Mateus static bending test ($\times 0.5$).

mation to measurement of modulus of elasticity (Bravery and Lavers, 1971). These workers have shown, however, that changes in modulus of elasticity were developed simultaneously with losses in weight in such small beams, yet more replicates and more careful measurement were necessary for the strength criterion. Sharp and Eggins (1968) used a relatively simple test in which bending strength was measured. The technique employed small samples stamped out of rotary-cut veneers but the source of material and preparation of the samples must inevitably lead to great variation within sets of replicates. Hopkins and Coldwell (1944), Kennedy and Ifju (1962), Wilson and Ifju (1965) and Smith (1970) are prominent among those who have used small scale tensile strength testing techniques to assess rates of decay in untreated and preservative-treated timber. Richardson (1968) has proposed a small-scale test which essentially relies upon the operator's ability to detect differences in tensile strength of decayed specimens when pulled between the fingers. Bravery and Grant (1971) described preliminary experiments in which the tensile testing techniques developed by Smith (1970) for soil burial tests, were applied to the assessment of preservative effectiveness against Basidiomycetes in monoculture (Fig. 8).

For precise results, all the methods employing strength criteria require careful sample preparation and comparatively large numbers of replicates because these criteria are particularly sensitive to variation in experimental technique and the inherent variability of wood, particularly wood undergoing decay. Samples have also to be saturated or conditioned to a given equilibrium moisture content before assessment and the methods of assessment themselves are frequently rather sophisticated and almost always time consuming.

Respirometric techniques such as those measuring carbon dioxide evolution (Klingström, 1965; Smith, 1967; Smith and Wilson, 1967) or oxygen consumption (Damaschke and Becker, 1965; Halabisky and Ifju, 1968) permit rapid evaluation of fungal attack but require sophisticated apparatus unsuitable for most routine studies (Fig. 12).

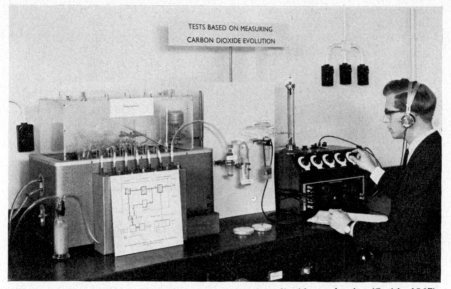

FIG. 12. Respirometric apparatus for measuring carbon dioxide production (Smith, 1967).

Conclusions

1 It is apparent that for most laboratory testing of preservative chemicals with monocultures of wood-destroying fungi, weight loss will remain the preferred standard criterion of attack. However there is a real need for simpler and more rapid techniques.

2 Continuing developmental research should aim to permit standardization of: (a) a more reliable initial screening test, (b) a test against blue-stain

in service, and (c) a small scale wood block test yielding results in less than half the time currently required.

3 Before it will be possible to predict preservative performance in service from laboratory test results, still further detailed studies are required on the complex of interrelated factors responsible for decay in treated wood.

References

ALLIOT, H. (1945). Methode d'essais des produits anticryptogamiques. *Bull. tech. Inst. nat. Bois*, No. 1.

ARMSTRONG, F. H. & SAVORY, J. G. (1959). The influence of fungal decay on the properties of timber. Effect of progressive decay by the soft rot fungus *Chaetomium globosum* on the strength of beech. *Holzforsch*, **13**, 84.

ASSOCIATION FRANCAIS DE NORMALISATION (1961). Protection du Bois. Methode d'essai des produits fungicides pour la protection des bois resineux de construction, utilisés dans les regions boreales. Standard No NFX 41-502: 1961 and NFX 41-512: 1961.

ASTM (American Society for Testing and Materials) (1961). Standard method of testing wood preservatives by laboratory soil-block cultures. ASTM Standard No D1413-61.

BAKER, J. M., MORGAN, J. W. W. & SAVORY, J. G. (1972). Wood protection research at the Forest Products Research Laboratory. *Rec. Ann. Conv. BWPA*, 183.

BRAVERY, A. F. (1968). Determining the tolerance of soft-rot fungi to wood preservatives: A comparison of test methods. *Mat. u. Org.*, **3**, 213.

BRAVERY, A. F. (1970). Preliminary observations on some effects of wood cell wall penetration by organic solvent type wood preservatives. *Int. Biodetn. Bull.*, **6**, 145.

BRAVERY, A. F. & GRANT, C. (1971). Preliminary investigations into the use of thin strip tensile strength tests for rapid evaluation of wood preservatives against Basidiomycetes. *Int. Biodetn Bull.*, **7**, 169.

BRAVERY, A. F. & LAVERS, G. M. (1971). Strength properties of decayed softwood measured on miniature test beams. *Int. Biodetn Bull.*, **7**, 129.

BRITISH STANDARDS INSTITUTION (1961). Methods of test for toxicity of wood preservatives to fungi. BS 838:1961.

BRITISH STANDARDS INSTITUTION (1968). Methods of test for fungal resistance of manufactured building materials. BS 1982:1968.

BUTIN, H. (1961). Ein neues Verfahren zur Bewertung der bläuewidrigen Wirksamkeit öliger Grundiermittel. *Holz als Roh und Werk*. **19**, 195.

CARTWRIGHT, K. ST G., CAMPBELL, W. G. & ARMSTRONG, F. H. (1936). Influence of fungal decay on the properties of timber. I The effect of progressive decay by *Polyporus hispidus* on the strength of English ash, *Fraxinus excelsior* L. *Proc. Roy. Soc. B.*, **120**, 76.

CARTWRIGHT, K. ST. G. & FINDLAY, W. P. K. (1958). Decay of timber and its prevention. 2nd Edn. HMSO: London.

COLLEY, R. H. (1953). The evaluation of wood preservatives. Part I. *Bell System Tech. J.*, **22**, 120.

CSERJESI, A. J. & ROFF, J. W. (1970). Accelerated laboratory test for evaluating the toxicity of fungicides for lumber. *Mat. Res. Stand.*, **10**, 18.

DAMASCHKE, K. & BECKER, G. (1965). Rythmen des Sauerstoffverbrauchs von Basidiomyceten. *Holz u. Organismen*, 275–290. Supplement to *Mat. u. Org.* (G. Becker and W. Liese, eds.). Duncker & Humbolt, Berlin.

DICKINSON, D. J. (1971). Disfigurement of decorative timbers by blue stain fungi. *Rec. Ann. Conv. BWPA*, 151.

DIN 52. 176:1961. Prüfung von Holzschutzmitteln-Mykologische Kurzprüfung, Klötzchen-Verfahren. *Deutscher Normenausschuss*.

DUNCAN, C. G. (1953). Soil block and agar block techniques for evaluation of oil-type wood preservatives: creosote, copper naphthenate and pentachlorophenol. *US Dept. Agric. for Path Special Release No 37*, Jan 1953.

DUNCAN, C. G. (1958). Studies on the methodology of soil-block testing. *FPL. Madison Rep.*, No. 2114.

FINDLAY, W. P. K. (1953). Influence of sample size on decay rate of wood in culture. *Timber Tech. and Machine Woodworking*, April 1953, 3 pp.

FINHOLT, R. W., WEEKS, M. & HATHAWAY, C. (1952). New theory on wood preservation. *Ind. and Eng. Chem.*, **44**, 101.

FLEROV, B. & POPOV, C. A. (1953). Methode zur Untersuchung der Wirkung von antiseptischen Mitteln auf holzzerstörende Pilze. *Angew. Bot.*, **15**, 386.

GIBSON, E. J. (1966). The role of laboratory testing in the evaluation of wood preservatives. *Wood*, June 1966, 3.

GRANT, C. (1973). The use of a reflectance method for estimating surface mould growth on chipboard. *Int. Biodetn Bull.*, **9**, 139.

HALABISKY, D. D. & IFJU, G. (1968). Use of respirometry for fast and accurate evaluation of wood preservatives. *Proc. AWPA*, **64**, 215.

HARTLEY, C. (1958). Evaluation of wood decay in experimental work. *US For. Prod. Lab. Rept.*, No. 2119.

HATFIELD, I., SHUMARO, R. S., VAUGHN, T. H. & HILL, E. F. (1950). The unseasoned wood disc method for evaluating sap-stain control chemicals. *Phytopath.*, **40**, 653.

HILDITCH, E. A. & HAMBLING, R. E. (1971). Wood preservative tests and their assessment. *Rec. Ann. Conv. BWPA*, 95.

HOPKINS, C. Y. & COLDWELL, B. B. (1944). Surface coatings for rot proofing wood. *Canad. Chem. and Proc. Ind. NRC*, No. 1256.

HUMPHREY, C. J. & FLEMING, R. M. (1915). The toxicity to fungi of various oils and salts particularly those used in wood preservation. *Bull. US Dept. Agric.*, No. 227.

KENNEDY, R. W. & IFJU, G. (1962). Application of microtensile testing in thin wood sections. *Tappi*, **45**, 725.

KIRK, H. & SCHULZ-DEWITZ, G. (1968). Elastizität und Schlagzähigkeit und Ihr Werserwert für die Wirksamkeit von Holzschutzmitteln gegenüber Ascomycetes, Fungi Imperfecti und Basidiomycetes. *Holztech.*, **9**, 249.

KLINGSTRÖM, A. (1965). Carbon dioxide as a measure of decay activity in wood blocks. *Studia Forestalia Suecida*, **26**, 1965.

LEVI, M. P. (1969). A rapid test for evaluating the fungicidal activity of potential wood preservatives. *J. Inst. Wood Sci.*, **23**, 45.

LEUTRITZ, J. (1946). A wood soil contact culture technique for laboratory study of wood destroying fungi, wood decay and wood preservation. *Bell Syst. tech. J.*, **25**, 102.

LIESE, J., NOWAK, A., PETERS, F. & RABANUS, A. (1935). Toximetrische Bestimmung von Holzkonservierungsmitteln. *Angew. Chem.*, **48**, 1.

MATEUS, T. J. E. (1954). Evaluation of the effectiveness of wood preservatives by a new method based on the measurement of deflections. *Portugal Min. of Public Works. Nat. Lab. Civil Eng. Publ.*, No. 48.

MATEUS, T. J. E. (1957). A mechanical test for studying wood preservatives. *Rec. Ann. Conv. BWPA*, 137.

NORDIC WOOD PRESERVATION COUNCIL. (1970). Testing of wood preservatives—Jordburk method—a soil block test with wood rotting Basidiomycetes. NWPC Standard No 1.4.1.1/70.

NORDIC WOOD PRESERVATION COUNCIL. (1970). Testing of wood preservatives—Mullåde method—a soil block test in unsterile soil. NWPC Standard No 1.4.1.2/70.

O'NEIL, L. A. & SKINNER, C. E. (1968). New laboratory test for assessing the resistance of paints to mould growth under tropical conditions. *IX Congress Fatipec*, Section 2, 12–14.

RABANUS, A. (1931). Über die Säure-Produktion von Pilzen und deren Einfluss auf die Wirkung von Holzschutzmitteln. *Mitt. d. Fachauss für Holzfragen b.V. deut. Ingen. und Deut. Forstverein*, **23**, 77.

RICHARDSON, B. A. (1968). A new technique for the comparative evaluation of some organo-metallic wood preservatives. *Int. Pest Control*, **10**, 14.

RHODES, F. H. & GARDNER, F. T. (1930). Comparative efficiencies of the components of creosote oil as preservatives for timber. *Industr. Engng. Chem.*, **22**, 167.

SAVORY, J. G. (1954a). Breakdown of timber by Ascomycetes and Fungi Imperfecti. *Ann. appl. Biol.*, **41**, 336.

SAVORY, J. G. (1954b). Damage to wood caused by micro-organisms. *J. appl. Bact.*, **17**, 213.

SAVORY, J. G. & BRAVERY, A. F. (1970a). Collaborative experiments in testing the toxicity of wood preservatives to soft rot fungi. *Mat. u. Org.*, **5**, 59.

SAVORY, J. G. & BRAVERY, A. F. (1970b). Observations on methods of determining the effectiveness of wood preservatives against soft rot fungi. *Mitt. dt. Ges. Holzforsch.*, **57**, 12.

SCHEFFER, T. C. (1936). Progressive effects of *Polyporus versicolor* on the physical and chemical properties of red gum sapwood. *US Dept. Agric. Tech. Bull.*, No. 527.

SCHMITZ, H. *et al.* (1930). A suggested toximetric method for wood preservatives. *Industr. Engng. Chem., Anal. Edn.*, **2**, 361.

SCHULZ, G. (1951). Ein mykologisches Verfahren zur Bewertung vorbeugender Schutzmittel gegen das Verblauen von Kiefernholz. *Angew. Bot.*, **26**, 1.

SCHULZ, G. (1952). Ein mykolgisches Verfahren zur Bewertung Fungicider Grundiermittel mit bläuewidriger Wirkung. *Holz als Roh und Werk.*, **10**, 353.

SHARP, R. F. & EGGINS, H. O. W. (1968). A rapid strength method for determining the biodeterioration of wood. *Int. Biodet. Bull.*, **4**, 63.

SMITH, D. N. R. (1965). The evaluation of wood preservatives. *Rec. Ann. Conv. BWPA*, 123.

SMITH, D. N. R. (1970). A possible method for the rapid evaluation of wood preservatives. *Mitt. dt. Ges. Holzforsch.*, **57**, 18.

SMITH, R. S. (1967). Carbon dioxide evolution as a measure of attack of wood by fungi and its application to testing wood preservatives and sapstain preventives. *Ann. appl. Biol.*, **59**, 473.

SMITH, R. S. & SAVORY, J. G. (1965). The toxicity of aged creosote to *Lentinus lepideus*. *J. Inst. Wood Sci.*, **13**, 31.

SMITH, R. S. & WILSON, W. (1967). Improved conductometric measurement of carbon dioxide. *Lab. Pract.*, **16**, 1377.

UNLIGIL, H. H. (1972). Tolerance of some Canadian strains of wood rotting fungi to wood preservatives. *For. Prod. J.*, **22**, 40.

WÄLCHLI, O. (1969). Die Prüfung der Wiederstandsfähigkeit von Holzschutz-mitteln gegen Moderfäule. *Schweizer. Arch. angew. Wiss. Tech.*, **35**, 73.

WILSON, J. W. & IFJU, G. (1965). Wood characteristics. VI Measuring density and strength properties of minute wood specimens. *Pulp Paper Res. Inst. Can. Tech. Rep.*, 423.

ZYCHA, H. (1964). Einwirkung einiger Moderfäule-Pilze auf Buchenholz. *Holz als Roh und Werk.*, **22**, 37.

Techniques for the Assay of Effects of Herbicides* on the Soil Microflora

E. GROSSBARD

ARC Weed Research Organization, Begbroke Hill, Yarnton, Oxford OX5 1PF, England

General Introduction

In the field of herbicide studies microbiological assay methods for the detection and estimation of the chemical present in a certain environment have not been developed to any extent. Nevertheless, algae have been used for bio-assay of herbicides since many are inhibitors of photosynthesis (S. J. L. Wright, see page 257). However, extensive microbiological studies have been made on the effect of herbicides on the soil microflora. It has been clearly established that herbicides exert an effect on the growth and activities of the soil microflora. These can be both inhibitory and stimulatory (Fletcher, 1966; Audus, 1970; Grossbard, 1972, 1973a).

At the Weed Research Organization (WRO) the evaluation of established and new herbicides includes an assessment of the responses by the soil microflora. A principal aim is to ascertain whether or not herbicides exert an adverse effect on soil fertility by virtue of any inhibitory effect on microbial activities. Herbicides are selective in their effect on different micro-organisms and their activities. Therefore several parameters must be tested simultaneously. The selection of valid criteria by which to assess the effect of herbicides, and of pesticides in general, presents a number of difficulties (Grossbard, 1970a, 1973a). This is associated in part with the difficulty, that soil microbiologists experience, to assess accurately the precise contribution of the soil microflora to soil fertility. Physiological groups such as nitrifying, nitrogen-fixing bacteria, cellulose decomposing organisms etc. are an exception. A further complication is concerned with techniques. Standard soil microbiological methods are frequently tedious and consequently costly. They do not necessarily reflect the real situation

---

* For the chemical names of the herbicides quoted see any recent issue of *Weed Abstracts* published by Commonwealth Agricultural Bureaux, Farnham Royal, Slough SL2 3BN, England.

in the field. In the evaluation of pesticides large numbers of chemicals have to be screened. In studies involving soil, whether the samples are taken from field or laboratory experiments, considerable replication is required, because of the heterogeneity of the soil *per se* and the uneven re-distribution of herbicide after application (Grossbard, 1970*b*). Furthermore, experiments should be carried out on a variety of soils since different soil characteristics will modify the behaviour of the herbicides and consequently their action. Since micro-organisms are prime agents in the degradation and/or detoxification of pesticides, important interactions between herbicides, their residues and their metabolites with the microflora and soil fauna will occur. Consequently, the microbial responses will change with time of incubation, thus requiring experiments of relatively long duration.

Extensive replication is needed, leading to the handling of large numbers of samples. On the other hand, only large differences between pesticide-treated and untreated control soils may be considered of interest because of the assumption that a suppression of microbial activity must be of a large order of magnitude in order to affect soil fertility. How far such a belief is justified cannot be judged by our present state of knowledge and it may be open to criticism. There is, therefore, a need for relatively simple and rapid techniques especially for those that would lend themselves to automation.

The methods described in the two parts of this paper refer to the effects of herbicides on symbiotic nitrogen fixation and on cellulose decomposition. These are two of several parameters at present used routinely in the microbiological assessment scheme at WRO. These activities are clearly and positively correlated with soil fertility.

It is not claimed that the techniques discussed here are the best and most precise that are known. Their choice was dictated by a requirement of speed, convenience, ease of manipulation and suitability for the limited facilities available while at the same time combining the greatest validity and accuracy possible within these limits.

PART I

The effect of herbicides on *Rhizobium trifolii* and symbiotic nitrogen fixation—E. GROSSBARD

Introduction

One of the principal contributions to soil nitrogen originates from the fixation of atmospheric nitrogen by bacteria especially those acting in symbiosis with legumes such as rhizobia (Nutman, 1963; Masterson and

Sherwood, 1970). Thus, any interference in this activity by agricultural chemicals applied either to the foliage of legumes or to the soil directly might reduce soil fertility. The nitrogen-fixing activity of rhizobium is inter-dependent on the active growth of its host. By definition herbicides are phytotoxic, thus in evaluating the effect of herbicides on rhizobia and their activities a distinction must be made between the action of the herbicides on the bacterium *per se* from that on the host, i.e. the legume, and finally on the symbiotic relationship. Some herbicides may severely inhibit the growth of many legumes while others such as MCPB, 2,4-DB (Fletcher *et al.*, 1957) and benazolin (Fryer and Makepeace, 1972) may be relatively innocuous to one or more leguminous crops. This is thus an instance where the microbiological assay must also involve a higher plant.

The effect of herbicides on nitrogen fixation may be concerned with:
(i) growth and activity of rhizobia—different species and strains reacting differently (Kaszubiak, 1966; Jensen, 1969);
(ii) changes induced by herbicides in the general soil microflora which may have repercussions on rhizobia;
(iii) growth of the host, and
(iv) the symbiotic relationship between rhizobia and legume.

The responses of leguminous crops to herbicides, including nodulation, is widely studied in field and pot experiments. Investigations in aseptic cultures of legumes (Fletcher, Dickenson and Raymond, 1956; Jensen, 1969) and in pure cultures of rhizobium (e.g. Kaszubiak, 1966; Grossbard, 1970c) allow more detailed investigations.

Techniques at W.R.O.

The present studies use pure cultures of both *Rhizobium trifolii* and white clover (*Trifolium repens* L). They are intended to test the ability of rhizobia that were pre-treated with herbicides, to infect subsequently the host plant and fix successfully atmospheric nitrogen.

In the first instance the bacterium is exposed to the herbicide for a short time to simulate the field situation where the chemical is applied to the soil shortly before emergence of the legume host. Long exposure experiments are designed to test the response of rhizobium in the absence of the host, simulating application in the course of growing non-leguminous crops. Here the bacteria may be exposed for long periods to gradually decreasing residues of a persistent herbicide. The survival, and the ability of the bacteria to infect the legume when it is eventually sown, and the "effectiveness" of fixing nitrogen is then examined.

The methods described here are based to a large extent on standard techniques as reviewed by Vincent (1970) with some modification devised specifically for the assay of herbicidal effects.

Bacterial culture

Rhizobium trifolii, strain 5, kindly supplied by Dr. P. Nutman from the culture collection at the Rothamsted Experimental Station, Harpenden, Herts, England. Some studies are carried out also on strain 1.

Media

A modification of the medium of Kleczkowska (1945 and pers. comm., 1969) is used. The standard calcium source of $CaCO_3$ or $CaCl_2$ (Vincent, 1970) is replaced by calcium gluconate to facilitate sedimentation of the cells by high speed centrifugation (9000 revs/min) at 10° for 15 min. Two media are used, one which serves as the standard and the other which is calcium-deficient—calcium gluconate is omitted and glass-distilled water is used. It has been found that the absence of calcium in the growth medium affects the response of *R. trifolii* to various herbicides. This is independent of the reaction of the medium (Grossbard, 1971, in press). The nutrients in the media were made up to eight times their normal concentration to give on dilution, and after addition of the appropriate herbicide, a final composition of (g/l):

(1) Mannitol	10·0
(2) $MgSO_4.7H_2O$	0·2
(3) K_2HPO_4	0·5
(4) NaCl	0·1
(5) Yeast extract (Oxoid or Difco)	0·2
(6) Calcium gluconate	0·5
(7) Ferric citrate	0·2
(8) Agar (Oxoid)	15·0
(9) De-ionized H_2O to	1 litre

In some experiments mannitol is replaced by glucose to decrease viscosity of the culture.

Ingredients 1, 2, 4 and 5 are dissolved in water, mixed and added to the molten agar. Ferric citrate and calcium gluconate are dissolved separately and then added to the agar, the mixture filtered, dispensed and autoclaved. The phosphate is sterilized separately and added aseptically to the sterile medium. The pH of the completed standard medium is 6·9 and that of the calcium-deficient is 7·0 after autoclaving and addition of the phosphate.

Herbicides

Technically-pure herbicides are used when other components of the commercial product are not likely to have specific effects; otherwise the formulated material is tested. As many herbicides have a very low solubility in water their sterilization presents difficulties. Therefore, it is necessary to make up the nutrients at high concentrations so that only small quantities of stock solution of the herbicides—still within the limits of solubility—need to be used to obtain the required concentration. The appropriate dilution will then give equal amounts of nutrients in treatments and controls. In our work at WRO we use either 25 or 50 ppm of herbicides in the growth medium. This is a very high concentration as compared with that likely to occur in the soil solution as a result of field usage.

The chemical is passed through a bacteriological sintered glass filter and an aliquot added aseptically to the sterile medium just before inoculation. This is the most reliable method. Many herbicides such as linuron are readily soluble in methanol or ethanol. The required amount of herbicide is dissolved in a small quantity of the solvent, left to stand overnight and, before use, diluted with sterile water within the limits of water solubility. Though some herbicides are moderately heat stable we have avoided sterilization by autoclaving because of lack of information on the effect of autoclaving on many herbicides. Intermittent sterilization by heating at 95° for 5 min on 3 consecutive days gave satisfactory results with, for instance Simazine (Grossbard, 1970b). Sterilization by irradiation with *gamma*-rays must be avoided as some herbicides, such as for instance linuron, are degraded to a considerable extent by 25 Mrad. Wherever possible, the author recommends sterilization through sintered glass filter. Seitz filters and Millipore membranes may be unsuitable because of adsorption of herbicides. It is advisable to check the final herbicide concentration by chemical analysis in the medium. Even after passage through sintered glass filters, unexplained losses in herbicide occurred at times.

Host

White Clover, S.700.

Procedure prior to inoculation of leguminous host

Rhizobium tifolii is grown in 50 ml of medium, with or without the appropriate herbicide, in 250 ml Erlenmeyer flasks in shake culture at 22° for either 4, 5 or 7 days, according to the experiment. Four replicates per treatment are set up. The bacterial suspension, before harvesting of cells, is used for turbidity readings, plate counts and inoculation of clover seedlings. Of the residue 20 ml is spun down at 9000 rev/min for 15 min in a refrigerated centrifuge at 10°. The harvested cells are re-suspended in 20 ml of phosphate buffer. One half, i.e. "unwashed cells", is left untreated. The other half is washed four times by re-suspending in phosphate buffer and mixing on a "Whirlimixer" and recentrifuging to remove herbicide residues. These are the "washed" cells. Controls are treated in the same way as are also cultures from calcium-deficient media, provided the bacteria had grown at all in the latter. Aliquots of each type of cells are used for respiration experiments and inoculation of clover seedlings.

Growth measurements

At the conclusion of the treatment period the following measurements of growth are made (Fig. 1):
 (i) Turbidity readings on an Eel absorptiometer. This is carried out in all experiments using the bacterial suspension.
 (ii) Plate counts are made for the enumeration of viable propagules on a standard medium with calcium gluconate and congo red (Vincent, 1970).
The Miles and Misra method can be used as well (Vincent, 1970) but we find that the standard dilution plate method gives a better definition of individual colonies. Viable counts are made usually after growth of the bacterium for up to 7 d in the presence of the herbicides under test.

Respiratory activity

This is measured on a Gilson differential respirometer (Gilson, 1963). The parameter is used either for screening experiments in the initial stages of evaluation or for studies involving growth of rhizobium in the presence of herbicides.

Preliminary tests involve the simultaneous assay of a series of herbicides. Cells are grown for 4 days in a standard medium without herbicide, harvested, washed 3 times to remove the nutrients and re-suspended in phosphate buffer. A standardized suspension of 2·5 ml is placed in the centre of a conventional Warburg flask as well as 0·5 ml of mannitol giving a final concentration of 200 ppm of the sugar. The herbicide is in the side arm in appropriate amounts to give the desired concentration between 25–70 ppm when tipped. Where the water solubility is low these concentrations cannot be reached accurately as the herbicide in the side arm may be in a suspension. Either a lower total concentration is accepted or the for-

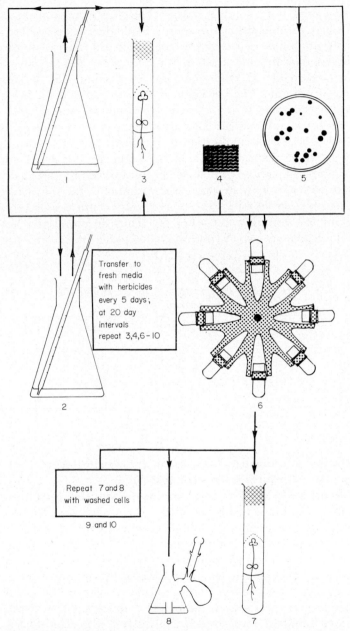

FIG. 1. Diagrammatic representations of various assays and of steps involved in repeated transfers during long term exposure of *R. trifolii* to herbicides. 1. Shake culture for 5 days. 2. Transfer five ml into fresh medium containing herbicide. 3. Inoculate sterile clover seedling with (1). 4. Read turbidity of (1). 5. Prepare plate counts of (1). 6. Harvest cells from a 20 ml aliquot of (1). 7. Re-suspend cells in buffer and inoculate clover. 8. Measure O_2 uptake of (6) in buffer. 9 and 10. Wash (6) four times in buffer and repeat (7) and (8).

mulated material, if it is soluble, is used. The effect of the solvent and other additives on oxygen uptake by rhizobium should then be assessed separately. If one solvent only is used, e.g. methanol as in the case of linuron, controls with appropriate concentration of methanol are also included. Oxygen uptake is measured at 25° while the flasks are shaking. The initial readings are taken for 90 min. The herbicide is then tipped and further readings continued for another 180 min. These screening experiments are useful to obtain rapidly an approximate forecast of the possible behaviour of a new compound. The second application of measurements of oxygen uptake is on cells grown in media, containing the herbicide, for 4 days or longer (Fig. 1). These are carried out on either "unwashed" or "washed" cells in phosphate buffer with either glucose or mannitol in the side arm which is tipped usually after 90 min. Results frequently confirm those obtained in the preliminary screening tests (Fig. 2). They do not always agree with growth measurement which indicates the need for several parameters (Table 1).

TABLE 1. The effect of dinoseb on oxygen uptake and counts of propagules of *Rhizobium trifolii*

Concentration (ppm)	$O_2\mu l/h$	Counts ($\times 10^8$)
0	42	46
2	82	35
25	75	21
50	62	18

Nitrogen fixation

At WRO this is estimated in terms of dry weight and content of total nitrogen of S 100 white clover seedlings.

The seeds, sterilized with sulphuric acid (Nutman, pers. comm.), are sown on water agar and incubated at 25°. When the radicle emerges and is *c*. 3 mm long, two seedlings of similar size are planted on seedling agar slopes (Vincent, 1970) in boiling tubes $2 \cdot 5 \times 20$ cm, and allowed to grow in the dark for 3 d. They are then placed, for the first few days in subdued, and then in normal light. After about 10 days growth one of the two seedlings is removed and the remaining inoculated with 0·5 ml of either the untreated bacterial suspension, or of the "unwashed" or "washed" cell suspensions in buffer. Two slopes are inoculated with cells of each of the 4 replicate flasks. Five ml of a solution containing the same nutrients as the seedling agar are added to each slope. In some experiments, especially in the late autumn and winter when growth is slow, 0·01% of ammonium

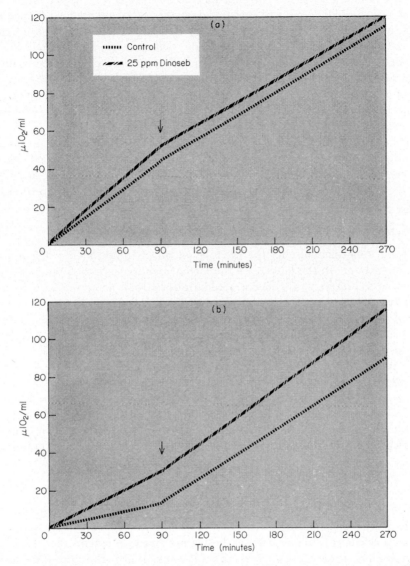

FIG. 2. A comparison of the effect of dinoseb at 25 ppm on washed cells of *Rhizobium trifolii*. (a) Previously grown in standard medium without herbicide and subsequently exposed for three hours—screening experiments. ↓ = herbicide tipped in one half of the flasks and water into the other (controls). (b) Grown in presence of herbicide for four days—growth experiments. ↓ = glucose tipped into all flasks. The source of carbon in the medium was glucose and not mannitol.

nitrate is added at the time of inoculation only. At WRO the seedlings are grown in a greenhouse with natural daylight, supplemented when necessary by light from fluorescent tubes to give a 14 h day length. Culture in an illuminated, constant temperature room led to symptoms of chlorosis though red light was supplemented and various trace elements supplied. Since many herbicides cause chlorosis it is of utmost importance not to confuse such symptoms with those that might be due to growth conditions. The seedling nutrient solution is replenished at intervals and the seedlings grown until they reach the cotton wool plug which may be between 7–10 weeks. During the growing period visual assessments are made at intervals of vigour, number of leaves, nodules, their position and colour etc. After harvest a careful count is made of number of leaves, roots and nodules. The seedlings are then dried overnight at 80° and the weight determined.

Determination of total nitrogen in clover seedlings

This is done manually by the micro-Kjeldahl method or on a Technicon AutoAnalyser. In the latter case the clover seedlings are digested as in the manual technique with concentrated sulphuric acid, using selenium as the catalyst. However, the total N in the digest is estimated on the Auto-Analyser colorimetrically as the indophenol-blue complex (Varley, 1966).

An alternative method of estimating nitrogen fixation is by the acetylene reduction technique (Hardy & Burns, 1968; Dart et al., 1972). We do not use this method since a gas chromatograph is not available to measure the ethylene produced by the action of nitrogenase on acetylene. This technique is, however, recommended where facilities are available for the sake of speed and accuracy. However, in the case of herbicides nitrogenase estimation alone may not always suffice when it is necessary to discriminate between the phytotoxic action of the herbicide and antibacterial potential.

Experiments over short periods

Initially *R. trifolii* was grown in the standard medium for the normal period of 4 days. When the experimentation was expanded to include a calcium-deficient medium it was found necessary to extend the period to 7 d because of the early and severe inhibitions of growth induced by some herbicides in the absence of calcium (Grossbard, in press).

Plate counts, measurements of turbidity and of oxygen uptake with "unwashed cells" as well as the inoculation of previously grown clover seedlings are carried out on the same day. Oxygen uptake by "washed cells" has to be measured the following day in view of the lengthy washing procedure (Fig. 1 of Table 2).

Experiments over long periods

These are based on a series of transfers of *R. trifolii* (Fig. 1). The cultures are treated as in the short period experiments. After 5 days growth, 5 ml of the bacterial suspension is transferred to fresh medium containing the herbicide under test in the same concentration as before and the transfers repeated every 5 days for up to 10 weeks. At 2, 5 and 10 week intervals—counted from the first date of inoculation—the cultures are sub-sampled, turbidity and oxygen uptake is measured and clover seedlings inoculated with the bacterial suspension, "unwashed" and "washed" cell suspensions. These seedlings are harvested after similar periods of growth and assessed as in short term experiments with which the results are compared.

The technique of repeated transfer is more suitable to study the effect of prolonged exposure than maintaining the same culture for a long time. The great viscosity of older cultures makes removal of traces of herbicides from the cells impracticable.

Discussion

Techniques described here are confined to pure culture methods of both the bacterium and its host, the latter after inoculation with pre-treated bacteria. The herbicide could not be applied simultaneously to both because in the concentrations used for rhizobium assays, many would be lethal to clover. Alternatively, a dose rate tolerated by clover has no effect on rhizobium (Grossbard, in press). The situation under study is primarily a model to forecast the response of bacteria exposed to herbicides in the absence of the host. For this reason such studies are now being initiated in soil using the enumeration techniques of rhizobia described by Vincent (1970). Growth measurements alone, such as counts, do not necessarily provide a reliable forecast on the outcome of the symbiotic relationship between rhizobium and its host. Results with some herbicides give lower figures for total, and even more so for viable counts, which is especially marked in the absence of a calcium source. Yet effectiveness of nitrogen fixation is not always affected provided traces of herbicides are not transferred on to the host via the inoculum. When sterile clover seedlings are inoculated with 0·5 ml of the culture previously grown in the presence of certain herbicides such as linuron, the seedlings die within 8–10 days. This occurs whether or not any decrease in growth of rhizobia has been recorded previously. About 0·5 ppm of herbicide may be present in the slope. If the inoculum consists of a suspension of "unwashed cells", the seedlings will form a plant of reduced weight and nitrogen content. This is probably due to the action of herbicides still adhering to the cells, though in smaller

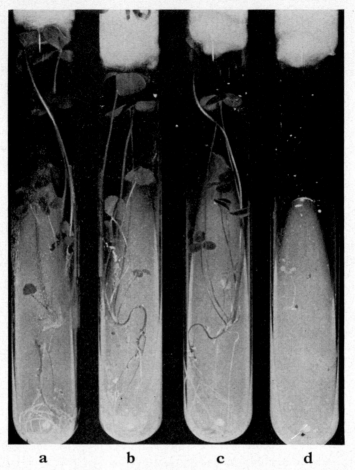

a   b   c   d

FIG. 3. The effect of small amounts of herbicides transferred to clover seedlings via the inoculum. Inoculum (a) Bacterial suspension from untreated culture; (b) "washed" cells from culture treated with linuron; (c) "unwashed" cells of (b); (d) Bacterial suspension of culture treated with linuron.

amounts as in the culture. If the bacteria are washed thoroughly to diminish further the trace amounts of herbicides, a normal plant develops (Fig. 3). This technique is a useful tool to illustrate the relative resistance of the bacterium as compared with the plant (Grossbard, 1970c), since the "effectiveness" of rhizobium seems not to be impaired after contact with the herbicides so far tested.

As the inhibition of growth of rhizobium itself, resulting from herbicide treatment, is not always reflected in results on nitrogen fixation one could argue that an assay of the legume alone would suffice. For practical

TABLE 2. A comparison of different parameters for the assay of effects of herbicides on *Rhizobium trifolii* in standard and calcium-deficient media containing 25 ppm of herbicides, % control

Herbicide (25 ppm)	Medium	Turbidity	Viable count ($\times 10^8$)	Respiration O_2 uptake (washed cells)	N_2-fixation in symbiosis with white clover					
					Dry weight			Total N		
					B. Susp.	Unwashed	Washed	B. Susp.	Unwashed	Washed
Linuron	Standard	91	90	115	0*	66	135	0	69	130
Asulam		63	96	80	89	–	104	114	–	100
Linuron	Calcium-Deficient	57	53	52	0*	84	96	0	112	86
Asulam		50	9	85	28	–	85	–	–	–

* Plants died within 10 days; B. Susp. = bacterial suspension

purposes it is this final step that matters, at least when dealing with a preliminary screening of large numbers of chemicals. However, the clover tests are too laborious and time-consuming, though assays involving nitrogenase measurements might help such investigations. It would be desirable to devise a simpler and faster technique than the legume test. Measurements on oxygen uptake tend to show more agreement with data on nitrogen fixation than do growth measurements (Table 2). However, not enough work has been done to justify the use of respiration as an "indicator activity" of herbicidal effects in this context.

For more detailed follow-up studies, plate counts and/or turbidity measurements should be retained. They provide valuable information on possible injury to the bacterium itself. These studies have helped to illustrate the effect of calcium in increasing the toxic effect of herbicides. Since both types of measurements show frequently a similar trend one of them may be dispensed with. Viable counts are probably a more reliable indicator of toxicity effects (Table 2) than turbidity determination, especially in an organism forming polysaccharides. Thus we retain plate counts for more important follow-up investigations especially on calcium-deficient media where older cultures are used. Whatever technique we have used, a reduction in cell count is the principal effect of the herbicides tested so far. Since only a few organisms are required for the infection of the legume, the inhibitory effect of herbicides on nitrogen fixation is, therefore, small or absent unless the herbicide causes genetical changes destroying "effectiveness".

PART II

The effect of herbicides on cellulose decomposition—
E. GROSSBARD AND G. I. WINGFIELD
ARC Weed Research Organization

Introduction

The degradation of this important polysaccharide is a microbial activity of major importance not only for soil fertility but also for the disposal of bulky organic matter in environments other than the soil, and finally as a contribution to CO_2 in the atmosphere.

The decomposition of cellulose, and of plant tissue in general, is of specific interest in relation to herbicide usage. Apart from the overall decomposition of organic matter in the soil, the disposal of vegetation i.e. plant remains, barley stubble, swards, etc. after treatment with herbicide is important in order to enable efficient re-sowing and establishment of a crop. This applies especially where minimal cultivation techniques are superseding conventional cultivations (Grossbard, Marsh and Wingfield, 1972). Furthermore, cellulolytic organisms, especially fungi, have been shown to be sensitive to herbicides at certain concentrations (Sobieszczański, 1969; Wilkinson and Lucas, 1969; Szegi, 1970). For these reasons extensive studies are carried out on cellulose decomposition in various laboratories including WRO.

Techniques of studying cellulose degradation and the micro-organisms involved are numerous (Tribe, 1957; Rybalkina and Kononenko, 1961; Eggins and Pugh, 1962; Reese, 1962; Webley and Duff, 1962; Went and de Jong, 1966; Latter, Cragg and Heal, 1967; Charpentier, 1968; Walsh and Stewart, 1969). Of these, some were selected and adapted at WRO for work with herbicides and others were developed. It is desirable to examine several parameters since cellulose decomposition can be measured from various aspects. As is the case with rhizobium, growth measurements do not always show the same trend as activity estimates.

A scheme is shown in Fig. 4 of assays that have been tested at WRO and compared with one another for relative efficiency and found to meet the requirements set out before (see page 225). They are arranged in a time sequence in as much as some can be used in a preliminary screening of a relatively large number of chemicals while further assays are employed in follow-up studies of a few selected herbicides.

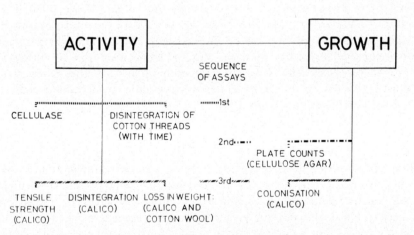

FIG. 4. Possible scheme for the assay of the effects of herbicides on microbial cellulose decomposition in the soil.

Materials

Soil

Sandy loam (pH 6·1) Available P 0·00125%; Organic C 1·6%, and Total N 0·17%.

Substrates

Cotton thread (50 gauge, J. and P. Coats Ltd.) is boiled twice in distilled water to remove any dressing.

Calico is a substrate that has been used for studies on microbial cellulose decomposition in soil by Latter, *et al.* (1967). Unbleached calico is boiled three times to remove any starch and other dressings and is well rinsed in tap water with a final two rinses in distilled water. It is then ironed to remove any creases and cut into strips, the long edge parallel with the selvedge.

Absorbent cotton wool is digested for 48 h in 270 ml concentrated HCl made up to 300 ml with distilled water (Hungate, 1950). Sufficient cotton wool is added to absorb all the acid. The digest is filtered in a Buchner funnel, well washed with distilled water to remove all traces of acid, dried and finally powdered in a disc mill (Glen Creston S.80).

Herbicides are used as commercial formulations.

Experimental procedures

Soil

Four kg of soil are placed in a polythene-lined metal drum on ball mill rollers (Grossbard and Marsh, 1974). While the drum is rotating herbicide and/or water are applied using an all-glass sprayer (A. Gallenkamp and Co. Ltd.) operated by compressed air (Fig. 5). Concentrations are either 500 ppm or the equivalent w/w concentration applied to the calico strips. Moisture content of the soil is adjusted to its field capacity and maintained at this level during incubation.

Substrates

The washed cotton threads are attached to a Dexion framework and herbicide applied as to the calico (Fig. 6). The washed calico is attached to a glass plate and herbicide applied using a laboratory pot sprayer embodying a "Teejet" fan nozzle moving at constant speed along a track over a spray bench (Fig. 7). The volume of liquid sprayed on to a unit area is determined by the size of the nozzle orifice, the distance below the nozzle, the speed of

FIG. 5. Application of herbicide to soil contained in a polythene-lined metal drum on ball mill rollers using an all-glass sprayer.

FIG. 6. Dexion frame-work to hold cotton threads in position when placed under pot sprayer.

FIG. 7. Laboratory pot sprayer for the application of herbicides.

travel and the spraying pressure. The concentration of herbicide (g/100 ml) required for a given dose is then calculated from the following equation:

$$\text{Concentration of active ingredient in g/100 ml} = \frac{x}{10y}$$

where x is the required dose in kg/ha and y is the deposition rate of the spray equipment in litre/ha. This, in turn, is multiplied by

$$\frac{100}{\% \text{ active ingredient}}$$

to give the amount of commercial product required. Dose may be converted into ppm on the basis of the dry weight of a given area of calico exposed to spray. Amounts used in this work are 3–4 times greater than in field applications.

The acid-treated cotton wool, as a dry powder, is sprayed with herbicide and water in the same manner as the soil. Loss of powder is reduced by covering the front of the drum with a polythene sheet. A small hole is made in the centre of the sheet thus allowing passage of the spray into the drum. The powder is first wetted with water, the herbicide then applied and the moisture content finally adjusted to 50% with a further volume of water. The concentration of herbicide on the powder is the equivalent to that on the calico.

Incubation techniques

Washed cotton threads are attached to the outside of a $15 \times 11 \times 7$ cm plastic box. The threads pass through small holes in the base of the box, and are kept taut over pullies with 50 g weights. The box is filled with soil adjusted to field moisture capacity, sealed in a polythene bag to prevent water loss and incubated at 24° (Fig. 8). Either soil or cotton thread may

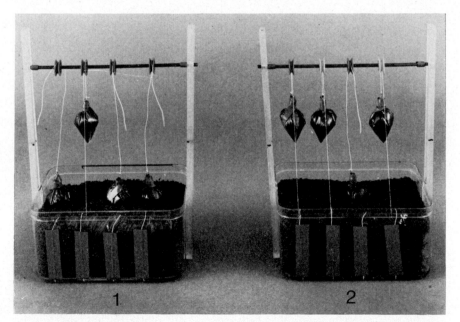

FIG. 8. Cotton thread decomposition. (1) control, and (2) soil treated with dalapon (500 ppm) after 4 days/24°.

be sprayed with herbicide. Alternatively, sub-samples of soil, previously treated with herbicide and then incubated for various periods of time, are used. The relative time taken for the threads to break serves as an approximate assessment of the effect of herbicide treatment on cellulose decomposition. The test is used as the first step in the screening programme.

Calico and powdered cotton wool are incubated on the surface or buried in soil. Calico strips (9·5 × 2·4 cm), to be buried, are attached by heat to weighed plastic cotton reels (Plastic Spools Ltd., Belle Vue Mills, Skipton, Yorkshire, B23 1RS) to aid the subsequent removal of the material. The total weight is recorded after overnight drying in a desiccator. Other strips are dried by the same method and placed on the soil surface.

About 10 g of powdered cotton wool are placed in weighed 10 × 10 cm squares of nylon mesh curtain material made into bags by sewing with nylon thread and the total wet weight then obtained. The dry weight is determined after overnight heating at 80°.

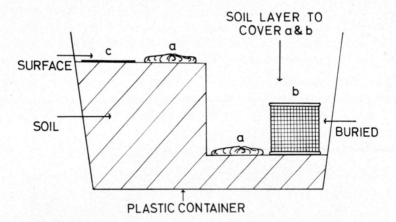

FIG. 9. Diagrammatic representation of arrangement of calico and powdered cotton wool on soil surface or when buried. (a) Cotton wool powder in nylon bag; (b) cotton reel with calico strip; (c) calico strip.

A layer of about 350 g of untreated or herbicide-sprayed soil is placed in a 22·5 × 14 × 8 cm plastic box on to which reels and bags, containing cellulose powder, are laid and susequently covered with 900 g soil to bury the cellulosic substrates. Bags and calico strips (9·5 × 2·4 cm) are laid on the soil surface (Fig. 9). The boxes are inserted in polythene bags which are filled with air, sealed and incubated at 23°. Aeration and adjustment of the soil moisture content is carried out every two weeks. Three boxes are used for each treatment and duplicate samples of both buried and surface cellulosic material removed after 4 and 8 weeks.

Parameters of activity

The loss in dry weight as a percentage of the weight before incubation is determined. Soil adhering to the calico strips or to the outside of the nylon bags containing powdered cotton wool is removed, as far as possible, with a camel hair brush. After overnight drying at 80° the bags are weighed and the dry weight of powder calculated (Fig. 10). Weight loss of calico strips is determined by a similar method except that drying is carried out in a desiccator at room temperature.

Tensile strength is calculated from breaking strain measured on 9.5×0.8

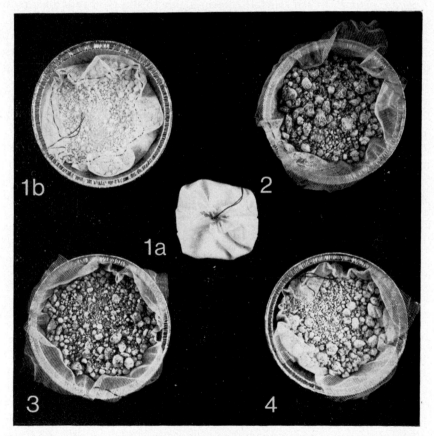

FIG. 10. Powdered, acid-treated cotton wool in nylon bags to determine loss in weight in soil treated with linuron after 8 weeks/24°. (1a) and (1b), before incubation; (2), after incubation in untreated soil (6·2% loss); (3), after incubation in linuron treated soil 50 ppm (6·1% loss); (4), after incubation in linuron treated soil 500 ppm (2·6% loss). Note marked discolouration in (2) and (3).

cm calico strips using a simple purpose-built tensile tester based on the sliding weight principle (after Latter, pers. comm.). Six determinations are made on each reel.

Disintegration is assessed on each strip using an arbitrary scoring system based on visual observation of destruction of the web and formation of holes in the fabric (Fig. 11).

<blockquote>
0   no disintegration
1–3 thin areas
4–6 small holes
7–9 large areas of fabric disintegrated
10   complete disintegration
</blockquote>

Cellulase activity in soil can be measured on freshly sprayed soil or on samples taken after incubation from the cotton thread or calico experiments. The technique of Benefield (1971) was modified by using an alternative method of estimating the reducing sugar derived from cellulase activity. In the original system, hydrogen peroxide formed during the dehydrogenation of glucose by glucose oxidase was catalysed by peroxidase at pH 7·0 to oxidize o-tolidine and a blue colour so formed. The modification is necessary since in our soils a brown colouration, possibly due to polyphenols, masks the blue colour.

Ten g soil and 120 mg powdered acid-treated cotton wool are incubated in 20 ml citrate/phosphate buffer (pH 5·0) containing $2 \times 10^{-4}M$ penicillin G, at 50° for 48 h. At the end of this period the incubation mixture is centrifuged at 3000 rev/min for 15 min. One ml supernatant, 1 ml distilled water and 2 ml copper reagent are boiled for 15 min and, after cooling, 2 ml arsenate reagent is added to develop the blue colour and the volume is then made up to 25 ml (Somogyi, 1945). The extinction of the colour is measured at 660 nm and reducing sugar concentration, as a direct measure of cellulase activity, determined from a glucose standard curve.

Parameters of growth

Plate counts of cellulolytic micro-organisms in soil are prepared from suspensions of herbicide-treated and control soils (1/10) in $\frac{1}{4}$-strength Ringer's solution. The suspensions are blended in a top drive macerator (Townson and Mercer Ltd.) and after appropriate dilution, 1 ml samples mixed with cellulose agar medium (Skinner, 1971; Hall and Grossbard, 1972). For the enumeration of fungal propagules, chlortetracycline hydrochloride ($30\mu g$/ml) is added to the medium to inhibit bacterial growth and, after incubation at 25° for 7 days, colonies producing areas of clearing of the medium are counted. For the enumeration of propagules of bacteria and of *Streptomyces* spp, fungal growth is suppressed by the addition of "Nysta-

FIG. 11. Disintegration of calico strip. (1), Calico before burial; (2), control after 8 weeks /24° on the soil surface. Note holes and thinning of fabric; scoring, 6, and weight loss, 54·2 %.

FIG. 12. Perspex scale over calico strip to measure area of colonization after 8 weeks/24° on the surface of untreated soil; 98 % colonization.

tin" (50 μg/ml) and cycloheximide (50 μg/ml) to the medium (Parkinson, Gray and Williams, 1971). Plates are incubated at 25° and colonies, showing an area of clearing of the medium, counted after 21 (bacteria) and 28 (*Streptomyces* spp) days.

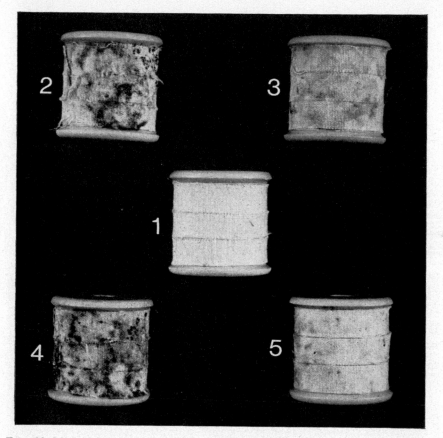

FIG. 13. Microbial colonization of Calico strips after herbicide application to either soil or Calico (4 weeks/24°). (1), before burial; (2), buried in untreated soil; (3) and (4), buried in soil treated with herbicides; (3), linuron (500 ppm); (4), paraquat (1330 ppm); (5), paraquat applied directly to calico (1330 ppm), soil un-treated.

Colonisation is assessed visually on calico strips. A piece of clear plastic divided into squares is placed on top and an estimate made as each square is about 1% of the total area of the strip. On the basis of colour, colonization by specific micro-organisms can also be assessed (Figs 12 and 13).
Isolations of individual species are made from small strands of calico and from larger pieces of the cloth which are shaken with ¼ strength Ringer's

solution. The suspensions are plated out on cellulose agar after appropriate dilution.

The Replidish technique is used to determine the effect of a range of concentrations of herbicides on cellulolytic fungal species. A method of multiple inoculation (Sneath, 1962; Goodfellow and Gray, 1966) has been adapted for this purpose at WRO. A Dyos Replidish of 25 compartments is filled with minimal agar (White and Downing, 1951) of the following composition (g/l): K_2HPO_4, 6·5; KH_2PO_4, 3·5; $(NH_4)_2SO_4$, 0·5; $CaCl_2$, 0·05; $MgSO_4.7H_2O$, 0·05; NaCl, 1·0; Agar (Oxoid No. 3), 12; Dist. H_2O to 1 litre; pH 7·0. The sterile herbicide is incorporated into the medium at a range of concentrations from 5–60 ppm and a square of sterilized calico placed on the surface of the agar in each compartment to serve as the sole source of carbon. Five replicates are made of each concentration and of the control. Five compartments are left uninoculated as a sterility check. A

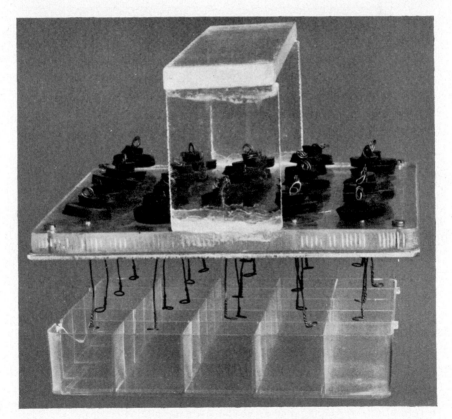

FIG. 14. Multipoint inoculator and Dyos Replidish.

cellulose agar slope is inoculated with the fungal isolate under test and after 3–4 days incubation a spore suspension prepared in $\frac{1}{4}$ strength Ringer's solution. Three ml of the suspension are then pipetted into each compartment of a second, empty Replidish leaving 5 sections unfilled. The Replidish with medium containing herbicide is subsequently inoculated with the spore suspension from the second Replidish using a multi-point inoculator (Quilt, 1972). This consists of a 6 mm Perspex block through which 25 holes are drilled to carry inoculating loops. A 1·5 mm aluminium plate is attached to the base and acts as a shield for the Perspex during flame sterilization of the loops (Fig. 14). After incubation at 25°, fungal growth, morphology and extent of sporulation in the compartments containing medium alone is compared with that in the herbicide-treated medium (Fig. 15). Results show that sporulation is reduced by linuron at 30 ppm and growth completely inhibited at 60 ppm.

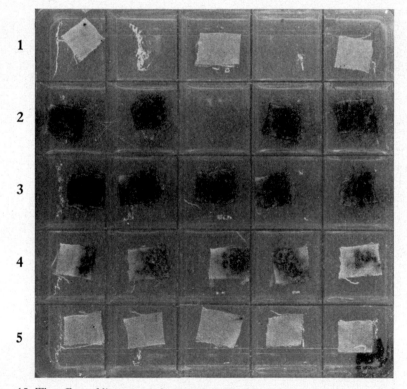

FIG. 15. The effect of linuron on the growth and sporulation of *Stachybotrys chartarum* (Ehrenb. ex Link) Hughes. Mineral medium with calico as sole carbon source. (1), not inoculated; (2), control; (3), linuron incorporated into medium at 5 ppm; (4), linuron incorporated into medium at 30 ppm; (5), linuron incorporated into medium at 60 ppm.

Discussion

The relative merits of various cotton substrates and the parameters to measure their degradation have been compared. The aim is to make a narrow selection of measurements to be used routinely in the assay of herbicidal effects on cellulose degradation.

Substrates

Cotton was chosen as substrate because its crystalline structure resembles that of the cellulose present in cell walls of higher plants; these studies being designed to serve as simulation of the degradation of vegetation in or on soil. At WRO we have adopted calico as the standard substrate for routine evaluation. It is preferable to powdered cotton wool because it is prepared more easily and degrades faster than does cotton wool (Table 3).

TABLE 3. *The relative decomposition of calico and powdered cotton wool sprayed with paraquat* and buried for 8 weeks in untreated soil*

Material	Treatment	
	Water	Paraquat
Calico	52·8**	42·1
Powdered cotton wool	12·3	7·4

* Sprayed at a rate equivalent to field application of 1·7 kg/ha which is 1300 ppm (w/w); ** % of original weight.

This is surprising in view of the similarity in the materials but it may be associated with the contact between substrate and soil. This is more intimate with the flat, thin calico than with cotton wool powder packed into nylon bags (Fig. 8). More measurements are possible on calico in addition to those which are made also on powdered cotton wool such as extent of distintegration, colonization and tensile strength.

Parameters of activity

Measurements of cellulolytic activity are of greater economic importance than are those of changes in growth and populations. For this reason they are more suitable for preliminary screening of a series of chemicals. Disintegration of cotton thread in terms of the time required for the weight to fall is a useful measurement. Also the determination of cellulase activity may lend itself well to this purpose, especially as it gives more precise quantitative data than the former. Both are non-specific gross measurements in as much as changes in size and composition of the cellulolytic microflora are not revealed. These should be the subject for detailed follow-up studies since they illustrate the effect of herbicides on the viability of the cellulolytic microflora which, in turn, influences activity. Tensile strength measurements may equally mask the true changes in the composition of the population. The material will break at the weakest spot which is the site colonised by the herbicide-resistant organisms while in fact large numbers of sensitive organisms may have been eliminated elsewhere.

Other factors leading to misinterpretation of the precise microbial participation may be due to mechanical disintegration of the substrate by soil animals, especially members of the mesofauna. This may facilitate subsequent microbial invasion. Furthermore, many soil animals also produce cellulase. However, Howard (1971) has devised a technique of irradiation with X-rays at a dose great enough to kill soil animals but not to affect micro-organisms. Alternatively, standard techniques of extracting soil animals especially the mesofauna, may help to reduce this form of error. Eggins and Lloyd (1968) recommend the use of glass fibre to protect cellulosic substrates in soil. This could be applied over calico strips to exclude a range of soil animals and would also avoid contamination by soil particles, another serious cause of error in weight loss determination. Application of this protective layer increases preparation time and may reduce decomposition. It is the contamination by soil that makes tensile strength measurement a more desirable parameter as compared with loss in weight as suggested by Latter (pers. comm.). The former, however is tedious and time-consuming unless sophisticated electronic apparatus is available.

In addition to the types of measurements described (Fig. 4) we have used oxygen uptake in studies on the decay of plant tissues sprayed with herbicides where the degradation of cellulose is also involved (Grossbard, *et al.*, 1972). Other parameters are based on autoradiography. ^{14}C labelled plant materials after spraying with herbicides are allowed to decompose on the soil surface with little disturbance. The fall in density of a series of autoradiographic images serves as an indicator of a gradual loss in carbon constituents with time of incubation (Grossbard, 1969, 1971, 1973b).

Parameters of growth

These are strictly related to the soil microflora. Measurements are made by plate counts, estimates of colonization of substrate such as calico, followed by more detailed studies of species composition. The disadvantages of plate counts for determining the number of viable microbial cells in soil are well known (Gray and Williams, 1971). In addition, the preparation, incubation and counting of plates is demanding both of time and manpower. However, interference by other agents such as animals obscuring inter-pretation of data is excluded. Moreover, gross changes in species composi-tion due to the effects of a herbicide can be detected by comparing the relative frequency of specific colonies by visual characterization based on morphology and colour. This may serve as a guide to types which are eliminated by the herbicide but present in controls and so enable the identification of sensitive species. The Replidish technique is a useful tool to follow up these observations by studies on individual isolates. The effects of herbicides at a range of concentrations against one soil isolate can be examined or of a single concentration against a large number of isolates under the same conditions. Although the costs of Replidishes are relatively high, the total expense of determining herbicidal effects on isolates is reduced if the multiple inoculation technique is used, because of the rapidity by which a large number of observations can be made as compared with conventional methods.

While results of different parameters may often show a similar trend (Table 4) with some herbicides, this is not always the case. In some instances a marked discrepancy may occur between data from, for instance, total counts and activity measurements and the response of individual species that are not reflected in either of the two former estimates.

Conclusions

No one single parameter is ideally suited to measure the effect of herbi-cides on either nitrogen fixation or cellulose decomposition. It is thus desirable to use a series of measurements and to consider the results together to detect a common trend. For preliminary screening tests, involving a large number of chemicals, the simplest activity measurement available should be selected. For practical purposes it is the final outcome, the sum total of one or several microbial activities that is of prime import-ance rather than the various factors that lead up to the ultimate effect. Most activity measurements are non-specific; they do not give precise indication of the action of herbicides on the viability of micro-organisms involved in the physiological activity under test. Thus growth and popula-tion studies should be the subject of detailed follow-up studies because in the final analysis they cannot be dissociated from parameters of activity.

Ultimately, the choice of criteria by which to assay effectively the action of herbicides or pesticides in general, lies with the research worker and facilities available. However, in view of the diversity of potential techniques and environmental conditions it is often difficult to compare results from

TABLE 4. *A comparison of different parameters for the assay of effects of herbicides on cellulose decomposition; calico and powdered cotton buried for 4 weeks in untreated soil*

| Treatment of substrate | % colonization* | Calico | | | Cotton wool Loss in weight as % original weight |
		Disintegration†* 0 (none)– 10 (total)	Tensile* strength (g/mm²)	Loss in weight as % original weight	
Water	86	2	72·9	31·6	6·4
Paraquat 1 300 ppm w/w	47	1	692·2	15·7***	4·1***

† Based on arbitrary scoring systems; * Statistical analysis not available, and ***, P < 0·001

sevei al laboratories even when dealing with the same pesticides. Thus a need exists for a system of guide lines to be developed that could help in the standardisation of both microbiological assay techniques for the detection of residual herbicides and their metabolites, and on the other hand of the effect of pesticides on the soil microflora and eventually soil fertility.

Acknowledgement

We wish to thank most sincerely Prof. J. D. Fryer and Dr K. Holly for valuable guidance and advice in this work; Miss B. E. M. Kinloch who assisted in the early stages of the studies on cellulose decomposition; Miss S. L. Giles for valuable collaboration; Miss C. J. Standell for technical assistance; Mr R. Harvey for the photographs and diagrams; Miss P. Latter for advice on the use of calico; Mr J. A. P. Marsh for analysis of nitrogen in clover; Dr P. A. Cawse for the irradiation of samples of herbicides and to Plastics Spools Ltd. for the gift of cotton reels. Several students on industrial training participated in the work on rhizobium and nitrogen fixation in recent years, i.e. Mrs G. Jessop and Miss J. P. Ricketts, both from Hatfield Polytechnic, Mrs T. A. Harvey, Liverpool Polytechnic and Mr J. Walshaw, London N.E. Polytechnic.

References

AUDUS, L. J. (1970). The action of herbicides and pesticides on the microflora. *Meded. Fac. Landbouw., Gent*, **35**, 465.

BENEFIELD, C. B. (1971). A rapid method for measuring cellulase activity in soils. *Soil. Biol. Biochem.*, **3**, 325.

CHARPENTIER, M. (1968). Dégradation de la cellulose dans le sol. Méchanismes enzymatiques. *Annls. Inst. Pasteur, Paris*, **115**, 497.

DART, P. J., DAY, J. M. & HARRIS, D. (1972). Assay of nitrogenase activity by acetylene reduction. Technical booklet on grain legume production. Vienna, FAO/IAEA.

EGGINS, H. O. W. & LLOYD, A. O. (1968). Cellulolytic fungi isolated by the screened substrate method. *Experimentia*, **24**, 749.

EGGINS, H. O. W. & PUGH, G. J. F. (1962). Isolation of cellulose decomposing fungi from soil. *Nature, Lond.*, **193**, 94.

FLETCHER, W. W. (1966). Herbicides and the bio-activity of the soil. *Landbouwk. Tijdschr.*, **78**, 274.

FLETCHER, W. W., DICKENSON, P. B. & RAYMOND, J. C. (1956). The effect of hormone herbicides on the growth and nodulation of *Trifolium repens sylvestre* in aseptic culture. *Phyton, B. Aires*, **7**, 121.

FLETCHER, W. W., DICKENSON, P. B., FORREST, J. D. & RAYMOND, J. C. (1957). The effect of soil applications of certain substituted phenoxy-acetic and phenoxybutyric acids on the growth and nodulation of *Trifolium repens sylvestre*. *Phyton, B. Aires*, **9**, 41.

FRYER, J. D. & MAKEPEACE, R. J. (Eds.) (1972). *Weed Control Handbook*, Vol. II, 7th Edn., pp. 1–30 and 420. Blackwell Scientific Publications, Oxford.

GILSON, W. E. (1963). Differential respirometer of simplified and improved design. *Science, N.Y.*, **141**, 531.

GOODFELLOW, M. & GRAY, T. R. G. (1966). A multipoint inoculation method for performing biochemical tests on bacteria. In *Identification Methods for Microbiologists, Part A*, (B. M. Gibbs and F. A. Skinner, eds.) p. 117. Academic Press, London.

GRAY, T. R. G. & WILLIAMS, S. T. (1971). *Soil Micro-organisms*, p. 73. Oliver and Boyd, Edinburgh.

GROSSBARD, E. (1969). A visual record of the decomposition of ^{14}C labelled fragments of grasses and rye added to the soil. *J. Soil Sci.*, **20**, 138.

GROSSBARD, E. (1970a). An appraisal of criteria by which to measure the effect of herbicides on the soil microflora, *Meded. Fac. Landbouw., Gent*, **35**, 515.

GROSSBARD, E. (1970b). The distribution of ^{14}C labelled simazine and atrazine before and after incubation detected by autoradiography of soil particles. *Meded. Fac. Landbouw., Gent*, **35**, 531.

GROSSBARD, E. (1970c). Effect of herbicides on the symbiotic relationship between *Rhizobium trifolii* and white clover. *Proc. Br. Grassl. Soc. White Clover Res. Symp.*, 1969, 47.

GROSSBARD, E. (1971). The purpose and significance of research on the microbial responses to herbicide application to soil. Paper read at: *Symp. of the Soc. of Chem. Ind. The effect of herbicides on the soil microflora and fauna.* London, 29th Nov., 1971.

GROSSBARD, E. & MARSH, J. A. P. (1974). The effect of some substituted urea herbicides on the soil microflora. *Pestic. Sci.* (in press).

GROSSBARD, E. (1972). Do herbicides affect the micro-organisms of the soil? *Rep. agric. Res. Coun. Weed Res. Orgn.*, **4**, 72.

GROSSBARD, E., MARSH, J. A. P. & WINGFIELD, G. I. (1972). The decay of residues of vegetation and of pure cellulose treated with aminotriazole and paraquat. *Proc. 11th Br. Weed Control. Conf.*, 673

GROSSBARD, E. (in press). The effect of herbicides on *Rhizobium trifolii*, with special reference to the role of calcium. In *Int. Symp. The interaction of herbicides, micro-organisms and plants.* Wrocław, 1973. Polish Journal of Soil Science.

GROSSBARD, E. (1973a). Problems of assessing the effect of pollutants on microbial activity. *Bull. Ecol. Res. Comm. Stockholm 17*, 457.

GROSSBARD, E. (1973b). Autoradiographic techniques in studies on the decomposition of plant residues labelled with ^{14}C. *Bull. Ecol. Res. Comm. Stockholm*, **17**, 279.

HALL, D. M. & GROSSBARD, E. (1972). The effects of grazing or cutting a perennial ryegrass and white clover sward on the microflora of the soil. *Soil Biol. Biochem.*, **4**, 199.

HARDY, R. W. F. & BURNS, R. C. (1968). Biological nitrogen fixation. *A. Rev. Biochem.*, **37**, 331.

HOWARD, P. J. A. (1971). Relationships between activity of organism and temperature and the computation of the annual respiration of micro-organisms decomposing leaf litter. *Proc. 4th Collq. Zool. Cttee Int. Soc. Soil Sci. I.N.R.A. Paris*, 197.

HUNGATE, R. E. (1950). The anaerobic mesophilic cellulolytic bacteria. *Bact. Rev.*, **14**, 1.

JENSEN, H. L. (1969). The effect of some herbicides on root nodule bacteria. *Tidsskr. Plavl.*, **73**, 309.

KASZUBIAK, H. (1966). The effect of herbicides on Rhizobium. I. Susceptibility of Rhizobium to herbicides. *Acta microbiol. pol.*, **15**, 357.

KLECZKOWSKA, J. (1945). The production of plaques by *Rhizobium* bacteriophage in poured plates and its value as counting method. *J. Bact.*, **50**, 71.

LATTER, P. M., CRAGG, J. B. & HEAL, O. W. (1967). Comparative studies on the microbiology of four moorland soils in the northern Pennines. *J. Ecol.*, **55**, 445.

MASTERSON, C. L. & SHERWOOD, M. (1970). Review. *Proc. Br. Grassl. Soc. White Clover Res. Symp.*, 1969, 11.

NUTMAN, P. S. (1963). Factors influencing the balance of mutual advantage in legume symbiosis. In *Symbiotic Associations* (P. S. Nutman and B. Mosse, eds), p. 51. University Press, Cambridge.

PARKINSON, D., GRAY, T. R. G. & WILLIAMS, S. T. (1971). *Methods for studying the ecology of soil micro-organisms.* I.B.P. Handbook No. 19, p. 106. Blackwell Scientific Publications, Oxford.

QUILT, P. (1972). *The effect of Carbyne on soil microorganisms.* Ph.D. Thesis, p. 42. University of Bath.

REESE, E. T. (1962). Advances in enzymic hydrolysis of cellulose and related materials. *Proc. Symp. Am. Chem. Soc. and U.S. Army research office, March 1972.* Pergamon Press, Oxford.

RYBALKINA, A. V. & KONONENKO, E. V. (1961). The microflora and nitrogen status of certain humus-peat soil. In *Micro-organisms and Organic Matter of Soils*, (M. M. Kononova, ed.), p. 3. Israel Program for Scientific Translations, Jerusalem 1970.

SKINNER, F. A. (1971). The isolation of soil clostridia. In *Isolation of Anaerobes*, (D. A. Shapton and R. G. Board, eds.), p. 68. Academic Press, London.

SNEATH, P. H. A. (1962). The construction of taxonomic groups. In *Microbial classification*, (F. Ainsworth and P. H. A. Sneath, eds.) p. 313. University Press, Cambridge.

SOBIESZCZAŃSKI, J. (1969). The effect of herbicides on soil microflora. V. growth and activity of cellulolytic microorganisms. *Acta microbiol. pol.*, **1 (18)**, 39.

SOMOGYI, M. (1945). A new reagent for the determination of sugars. *J. Biol. Chem.*, **1**, 61.

SZEGI, J. (1970). Effect of some herbicides on the growth of cellulose decomposing microscopic fungi. *Meded. Fac. Landbouwwetensch.*, *Gent*, **35**, 559.

TRIBE, H. T. (1957). Ecology of micro-organisms in soils as observed during their development upon buried cellulose film. *Symp. Soc. Gen. Microb.*, **7**, 287.

VARLEY, J. A. (1966). Automatic methods for the determination of nitrogen, phosphorus and potassium in plant material. *Analyst*, **91**, 119.

VINCENT, J. M. (1970). *A Manual for the Practical Study of Root-nodule Bacteria.* (I.B.P. Handbook No. 15), Blackwell Scientific Publications, Oxford.

WALSH, J. H. & STEWART, C. S. (1969). A simple method for the assay of the cellulolytic activity of fungi. *Int. Biodetn. Bull.*, **5**, 15.

WEBLEY, D. M. & DUFF, R. B. (1962). A technique for investigating localized microbial development in soils. *Nature, Lond.*, **194**, 364.

WENT, J. C. & DE JONG, F. (1966). Decomposition of cellulose in soils. *Antonie van Leeuwenhock*, **32**, 39.

WHITE, W. L. & DOWNING, M. H. (1951). *Coccospora agricola* Goddard its specific status, relationships and cellulolytic activities. *Mycologia*, **43**, 645.

WILKINSON, V. & LUCAS, R. L. (1969). Effects of herbicides on the growth of soil fungi. *New Phytol.*, **68**, 709.

WRIGHT, S. J. L. (1974). The use of micro algae for the assay of herbicides; this volume, page 257.

Use of Micro-Algae for the Assay of Herbicides

S. J. L. WRIGHT

*School of Biological Sciences, University of Bath,
Bath BA2 7AY, Avon, England*

Chemical methods for pesticide quantification, which may involve lengthy extractions and costly apparatus, are often used in studies of persistence, degradation and pollution. Such methods, however, do not necessarily distinguish between the pesticides and any structurally related but less toxic derivatives. Pesticide toxicity *per se* is therefore determined biologically. Herbicides are conventionally assayed by measuring parameters of growth or vigour in sensitive plant species grown on herbicide-treated soil, sand or filter-paper. Though sensitive, such assays are often lengthy and demanding in laboratory space and manpower.

Since in practice at least part of most pesticide applications reaches the soil and soil drainage water, it is important that the effects of such compounds on non-target organisms are investigated. Many algal species inhabit the soil and associated environments and though phylogenetically separated, the eucaryotic algae especially have physiological and cytological characteristics in common with the higher plants. Herbicides may reasonably be expected to act upon eucaryotic algae in a manner similar to that in which they affect the cells of susceptible plants; indeed algae have been used in studies relating to the mode of action of herbicides.

Before considering some examples of the use of algae for herbicidal assay it will be appropriate to cite some investigations which reflect the growing concern that soil and aquatic micro-algae may be vulnerable to the effects of pesticides.

Effects of Pesticides on Micro-Algae

The action of many phytotoxic compounds on algae has been investigated *in vitro* and several aspects of integrity and activity of the organisms have been examined. Among the reported effects are: growth inhibition (Geoghegan, 1957; Gramlich and Frans, 1964; Sikka and Pramer, 1968; Zweig, Hitt and McMahon, 1968; Loeppky and Tweedy, 1969; Ashton,

Bisalputra and Risley, 1966; Kirkwood and Fletcher, 1970; Wright, 1972); inhibition of photosynthetic oxygen evolution (Sikka and Pramer, 1968; Zweig et al., 1968); reduction in cell protein (Sikka and Pramer, 1968); reduced chlorophyll content (Ashton et al., 1966; Sikka and Pramer, 1968; Zweig et al., 1968); reduced phosphate uptake (Kirkwood and Fletcher, 1970), and reduced ATP levels (St. John, 1971). The unicellular green alga Chlorella was used by all the foregoing authors, while Loeppky and Tweedy (1969) and Kirkwood and Fletcher (1970) additionally included species of Chlamydomonas and Stichococcus and Chlamydomonas respectively. Maloney (1958) studied the inhibition of 33 algal species (green, blue-green and diatoms) and Venkataraman and Rajyalakshmi (1971) examined a range of pesticides with 28 strains of blue-green algae. Arvick, Willson and Darlington (1971) observed that although some soil algae were inhibited in vitro by high concentrations of a mixture of two herbicides, there was no detectable long-term change in the algal flora of treated soil. There have been relatively few reports on the effects of pesticides on marine unicellular algae. However, Walsh (1972) evaluated a wide range of herbicide types for their effects on the growth and O_2-evolution in vitro of pure cultures of unicellular marine chlorophyte and chrysophyte algae.

Algal Bio-Assay of Herbicides

There are clearly several aspects of algal growth and physiology which may be quantified in relation to the influence of inhibitors or stimuli. When selecting an alga for assay use it is important that the organism should not only respond sensitively to the factor under test but for added advantage should grow rapidly under easily reproduced conditions. It is also possible that some types of assay experiment would require the use of algae truly representative of the environmental algae, e.g. assessing toxicity and pollutant effects. It is noteworthy that while microbiological assays usually involve the use of specific strains of individual species, Tchan (1959) described a method for assaying available plant nutrients in soil using a mixture of soil algae.

In the method of Atkins and Tchan (1967) for determination of atrazine in soil, herbicide-treated soil was placed in cellophane containers and immersed in algal cultures for several days. Low concentrations of diuron were detected in aqueous extracts of soil using a 6 day Chlorella assay (Addison and Bardsley, 1968). The assay was quicker than an oat bio-assay and slower than a less-specific chemical method. Kratky and Warren (1971a) devised a simple and sensitive assay for photosynthetic inhibitors based on chlorophyll formation in Chlorella pyrenoidosa. The assay was particularly suited for leaching and residue studies and with results obtainable in 18–36 h was quicker than other *Chlorella* assays. The reduction in time was probably attributable to the use of a mixture of 3 % CO_2 and 97 % air for aeration of the cultures. Kratky and Warren (1971b) used the same assay to test 42 herbicides of different types and mode of action. The *Chlorella* assay was particularly sensitive to the photosynthetic and respiratory inhibitors and detected some compounds which plant root and shoot assays failed to detect. Several algae were evaluated for assay of urea herbicides (Pillay and Tchan, 1972) and an alga termed "NMI" selected for use in a liquid culture technique. Maximal inhibition was obtained in 3 days and the assay was more sensitive when the inoculum density was reduced. The same authors also described a method in which algae were inoculated on to paper discs incubated in contact with herbicide-treated soil and growth assessed visually after 30 h. Though less accurate and sensitive than other assays, the simplicity of the paper-disc technique favoured its use for rapid screening of soil suspected of containing high or low levels of herbicide.

Studies of movement in soil are important in consideration of the fate and behaviour of pesticides. Helling and Turner (1968) introduced soil thin-layer chromatography (TLC) of radio-labelled pesticides to investigate mobility. Labelled pesticides, when available, are expensive and the alternative method of detection was the relatively tedious chemical extraction and analysis of discrete zones from chromatograms. The use of algae for pesticide detection on soil TLC was subsequently proposed (Helling, Kaufman and Dieter, 1971). Pesticides were applied to soil TLC plates, which were then leached with water. *Chlorella* cells dispersed in agar were sprayed on to the plates giving the equivalent of an algal cell monolayer and after illumination under humid conditions for 24–48 h the position of pesticides was indicated by the appearance of clear areas in the algal growth. The technique was rapid, comparable in sensitivity to the use of labelled pesticides, detected several commercially important herbicides and was adaptable for detection of some fungicides and insecticides.

Whilst most of the techniques for algal bio-assay essentially involve culture of the organisms in mineral media under illumination, a variety of conditions have been specified. Culture vessels range from test-tubes to Erlenmeyer flasks with aeration, if provided, by agitation or bubbling mixtures of CO_2 and air. Temperatures range from 22° to 32° and illumination from greenhouse daylight to continuous light or set light periods in growth chambers. Considering the number of algal types, green and blue-green, inhabiting soil and aquatic environments there appears to be a wide choice of assay organisms. In fact few species have been used and there is a clear preference for unicellular green algae of the *Chlorella* type. *Chlorella* is in many respects an attractive organism for experimentation especially to those not trained in the methods of phycology. The ready availability from culture collections and research laboratories of a wide range of *Chlorella* strains which are easily propagated *in vitro* in illuminated, aerated mineral salts media has undoubtedly contributed to their popularity. Though not necessarily truly representative of the green algae, many of which are colonial or filamentous, the genus is none the less ubiquitous and typically found on soil and in water. *Chlorella* growth in liquid culture is readily measured turbidimetrically and the organism is suitable for studies on photosynthesis, respiration, chlorophyll content, metabolism and cytology. Even so it is not unlikely that other algae may be more sensitive to some herbicides, a view which is supported by some of our recent tests on several blue-green algae.

In the next section a detailed description will be given of two bio-assay techniques we have developed and which were demonstrated at the recent meeting. These are: (1) a tube method (liquid culture), and (2) an agar plate method.

Tube method (liquid culture)

Organism

Chlorella 211/8h, formerly known as *Chlorella pyrenoidosa* strain Emerson 3, was obtained from the Culture Centre of Algae and Protozoa, 36 Storey's Way, Cambridge, England.

Growth medium and culture conditions

A modified Knop's solution (Samejima and Myers, 1958) is used, containing (g/l): KNO_3, 1·25; KH_2PO_4, 1·25; $MgSO_4.7H_2O$, 2·5; $Fe_2(SO_4)_3$, 0·004, and sodium citrate, 0·3. To this is added 1 ml "microelements" solution containing (g/l): H_3BO_3, 2·9; $MnCl_2.4H_2O$, 1·8; $ZnSO_4.7H_2O$, 0·22; MoO_3, 0·018, and $CuSO_4.5H_2O$, 0·08. The complete medium is adjusted to pH 6·0 and autoclaved.

The algae are cultured in 20 ml medium in large glass test tubes (150×25 mm) each with an aeration tube (230×3.5 mm) fitted through a cotton-wool plug (Fig. 1). The culture tubes are modified to have a conical base to minimize the tendency for cells to sediment during incubation. Rubber bungs with a small groove cut along the edge for air-venting may be used in place of cotton-wool and have the advantage of supporting the aeration tubes more firmly. For a standard assay, the herbicide is dissolved in the growth medium (for procedure with relatively insoluble compounds see Clark and Wright, 1970) and the solution sterilized by membrane filtration.

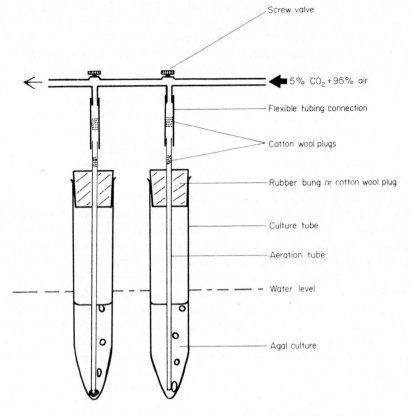

Screw valve

5% CO_2 + 95% air

Flexible tubing connection

Cotton wool plugs

Rubber bung or cotton wool plug

Culture tube

Aeration tube

Water level

Agal culture

FIG. 1. Assembly of tubes for culture of algae in liquid medium.

Appropriate concentration ranges are prepared by further dilution with sterile growth medium working to 19·5 ml volume and then inoculated with 0·5 ml cell suspension. The inoculum can be suitably standardized by diluting log phase cells, using growth medium, to an optical density value of 40 at 550 nm (EEL Spectra, Evans Electroselenium Ltd., Halstead, Essex, England), corresponding to approximately 0·55 mg/ml, dry weight of cells. After inoculation the tubes have a very faint green colouration. The tubes are supported in a Perspex rack and incubated in a transparent Perspex water bath (26×26×63 cm) thermostatically maintained at 25° using a heater/stirrer and cooling coil. Addition of a few drops of concentrated hypochlorite reduces the tendency for the water to become cloudy. Continuous illumination is provided by twin 80 W white fluorescent tubes arranged along opposite sides of the bath. Light intensity at the culture tubes is then 4,300 lux according to our measurements with a

distinct ranges and *Chlorella* was particularly sensitive to barban, fenuron and propanil. However, the ED_{50} values for propham and chlorpropham clearly indicate that this assay was less sensitive than a barley root assay for these compounds (Clark and Wright, 1970). Aniline compounds, which are formed during microbial metabolism of phenylamide herbicides (Wright, 1971), were far less inhibitory to *Chlorella* than the parent herbicides. The assay was adapted for monitoring toxicity during growth of a propham-degrading *Arthrobacter* sp. in propham-mineral salts medium. Decline in supernatant toxicity determined in this way closely paralleled the loss of propham as measured by uv absorption. According to the tube assay, over 90% of applied asulam at doses up to 5 ppm was recovered in aqueous extracts of treated soil, recovery falling to 80% at doses up to 10 ppm. The extracts, prepared by vigorously shaking samples of treated soil in growth medium for 1 h, were clarified by centrifugation and filtration prior to inoculation with *Chlorella*. Extracts of non-treated soil were used for control algal growth. Growth of blue-green algae in medium "D" of Kratz and Myers (1955) was inhibited by very low levels of propanil, ED_{50} values using *Gloeocapsa apicola* and *Nostoc muscorum* were 0·0025 ppm and 0·074 ppm respectively. The suitability of these and other blue-green algae for herbicide assay is currently being investigated.

Agar plate method

This method relies on the diffusion of herbicide from a suitable "reservoir" into an agar medium seeded with the algae. Growth of the algae on the medium surface is inhibited where the herbicide concentration is sufficiently high and the extent of inhibition indicated by the size of the consequent clear zone. The method is analogous to the antibiotic plate assays.

Procedure

We have used *Chlorella* 211/8 h and the medium is as described for the tube method except that 0·05 g ethylene diamine tetra acetic acid (EDTA) replaces citrate, to reduce contaminant bacterial growth, and is solidified with 1% (w/v) Ion Agar No. 2. (Oxoid). Fifteen millilitres of medium are dispensed in standard size Petri dishes and the agar surface is dried thoroughly. An aerosol spray ("Sprayit", Hewson Sturman Ltd., Bushey Heath, Herts), is used to uniformly inoculate the agar surface with a suspension of *Chlorella* cells taken from a 2 day liquid culture. The agar surface absorbs the inoculum so that when dry it has only a very faint green colour. The herbicide is dissolved in an appropriate volatile solvent

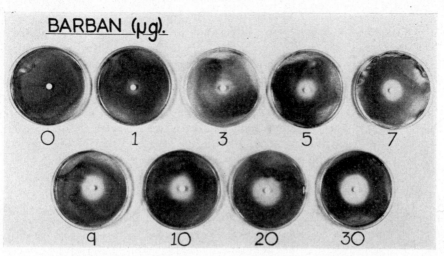

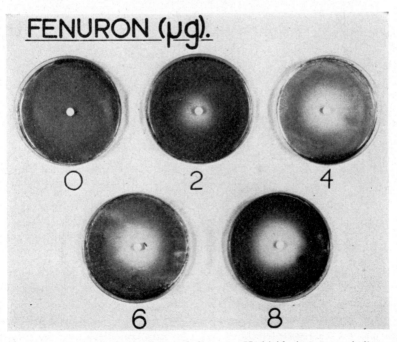

FIG. 2. Agar plate assay for barban and fenuron. Herbicide impregnated discs were placed on agar plates uniformly seeded with *Chlorella*. Photographs were taken after 3 days illumination.

(e.g. ethanol, acetone) and known amounts applied to 6·0 mm diam paper discs (Antibiotic Assay Discs, Whatman) using calibrated micro-pipettes (Drummond "Microcaps", Shandon Scientific Co. Ltd., London). Samples are added to the discs in 5 or 10 μl amounts and following evaporation of the solvent further additions may be made as required. Solvent only is applied to control discs. When dry, each herbicide-impregnated disc is placed on the dry surface of a *Chlorella* seeded plate. Plates are incubated at room temperature on a sheet of white paper 15 cm beneath twin 65/80 W white fluorescent strip lights. A thick film of algal growth develops in 2 to 3 days and clear areas of growth inhibition surround the herbicide discs. The diameter of each zone is measured, taking a mean of four determinations per zone.

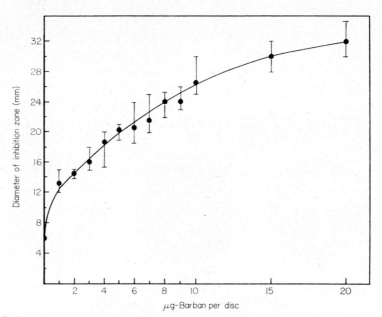

FIG. 3. Agar plate assay. Standard assay curve for barban using *Chlorella* 211/8 h (C.C.A.P.) Mean values are plotted together with limits from triplicate tests for each concentration in each of two separate experiments.

We have tested several herbicides and some algicides by this method. The assay is sensitive in most cases, size of inhibition zone relating to herbicide concentration (Figs 2 and 3), and detects the variation in potency among different herbicides. However, strict comparisons between herbicides cannot be made without taking into consideration the possible differences in diffusion properties. In some cases (e.g. fenuron) the zones are not bounded by a clear margin (Fig. 2) and are less readily quantified.

Though the choice of zone boundary in such cases becomes somewhat subjective, consistency can be achieved if the same operator examines the plates and works to a set standard. The assay detects fenuron and barban at levels as low as 1 μg per disc and is even more sensitive to some other herbicides. Notable in this respect were monuron and "Bladex" where a linear response was obtained between 0·4 and 1·0 μg and 0·1 and 1·0 μg respectively. The technique is probably better suited than the tube assay for determination of herbicides in soil. Herbicides are more readily extracted from soil using solvents and since the solvent evaporates from the assay disc before the disc is placed on the medium the algae are not affected (see control discs in Fig. 2). The foregoing details constitute a preliminary description of the plate assay method and current work on the use of other algae (green and blue-green), assay of herbicide residues in soil and application in toxicity testing will be reported elsewhere.

The agar plate assay is simple, inexpensive and gives results in 2 to 3 days. Having these advantages the technique might be useful in the primary screening of compounds for herbicidal activity, especially if more than one algal species is used. Compounds selected on this basis could then be evaluated more critically by assays using higher plants.

Maintenance of cultures

Stock cultures are grown at room temperature on agar slopes of the growth medium in test tubes or Universal bottles placed near a fluorescent light and sub-cultured fortnightly. Alternatively, cultures are maintained in aerated liquid medium and sub-cultured at 5 d intervals. Cultures may be stored at 5° for several months.

Conclusions

With both techniques described here results are quickly obtained, little laboratory space is required and the number of organisms in each experiment is large enough to offset individual variation in response which is more apparent in assays using plants. The *Chlorella* and other algae so far tested respond to a variety of herbicides, suggesting that relatively few algal species might be used to assay a wide range of phytotoxic compounds. It is also possible that some insecticides and fungicides could be assayed using algae. As with all microbiological assays it is important that the assay organism is genetically stable and this should be checked periodically with herbicide standards. Addison and Bardsley (1968) emphasized that all steps taken in the determination of unknowns should coincide with the procedure used in constructing standard curves and suggested the inclusion of a portion of the standard curve when running unknowns.

Apart from the suggestion that the *Chlorella* plate assay in particular may be useful in primary screening and toxicity testing, the methods described may be adapted for a number of experimental purposes and investigations including: detection and quantification of herbicide residues in soil, water or plants; toxicity tests on herbicide breakdown products; mode of action studies; the action of pesticides and other toxicants and pollutants on micro-algae; leaching of pesticides; bio-location of pesticides on chromatograms; bio-assay of soil fertility factors and evaluation of algicides.

Acknowledgements

The author gratefully acknowledges the technical assistance of Mrs Angela Forey and Mr Philip Walker and the financial support of the Agricultural Research Council.

References

ADDISON, D. A. & BARDSLEY, C. E. (1968). *Chlorella vulgaris* assay of the activity of soil herbicides. *Weed Sci.*, **16**, 427.

ARVICK, J. H., WILLSON, D. L. & DARLINGTON, L. C. (1971). Response of soil algae to picloram—2,4-D mixtures. *Weed Sci.*, **19**, 276.

ASHTON, F. M., BISALPUTRA, T. & RISLEY, E. B. (1966). Effect of atrazine on *Chlorella vulgaris*. *Am. J. Bot.*, **53**, 217.

ATKINS, C. A. & TCHAN, Y. T. (1967). Study of soil algae. VI. Bioassay of atrazine and the prediction of its toxicity in soils using an algal growth method. *Pl. Soil*, **27**, 432.

CLARK, C. G. & WRIGHT, S. J. L. (1970). Detoxication of isopropyl N-phenyl-carbamate (IPC) and isopropyl N-3-chlorophenylcarbamate (CIPC) in soil, and isolation of IPC-metabolizing bacteria. *Soil Biol. Biochem.*, **2**, 19.

GEOGHEGAN, M. J. (1957). The effect of some substituted methylureas on the respiration of *Chlorella vulgaris* var. *virids*. *New Phytol.*, **56**, 71.

GRAMLICH, J. V. & FRANS, R. E. (1964). Kinetics of *Chlorella* inhibition by herbicides. *Weeds*, **12**, 184.

HELLING, C. S., KAUFMAN, D. D. & DIETER, C. T. (1971). Algae bioassay detection of pesticide mobility in soils. *Weed Sci.*, **19**, 685.

HELLING, C. S. & TURNER, B. C. (1968). Pesticide mobility: determination by soil thin-layer chromatography. *Science, N. Y.*, **162**, 562.

KIRKWOOD, R. C. & FLETCHER, W. W. (1970). Factors influencing the herbicidal efficiency of MCPA and MCPB in three species of micro-algae. *Weed Res.*, **10**, 3.

KRATKY, B. A. & WARREN, G. F. (1971a). A rapid bioassay for photosynthetic and respiratory inhibitors. *Weed Sci.*, **19**, 658.

KRATKY, B. A. & WARREN, G. F. (1971b). The use of three simple, rapid bio-assays on forty-two herbicides. *Weed Res.*, **11**, 257.

KRATZ, W. A. & MYERS, J. (1955). Nutrition and growth of several blue-green algae. *Am. J. Bot.*, **42**, 282.

LOEPPKY, C. & TWEEDY, B. G. (1969). Effects of selected herbicides upon growth of soil algae. *Weed Sci.*, **17**, 110.

MALONEY, T. E. (1958). Control of algae with chlorophenyl dimethyl urea. *J. Am. Wat. Wks. Ass.*, **50**, 417.

PILLAY, A. R. & TCHAN, Y. T. (1972). Study of soil algae. VII. Adsorption of herbicides in soil and prediction of their rate of application by algal methods. *Pl. Soil*, **36**, 571.

SAMEJIMA, H. & MYERS, J. (1958). On the heterotrophic growth of *Chlorella pyrenoidosa*. *J. gen. Microbiol.*, **18**, 107.

SIKKA, H. C. & PRAMER, D. (1968). Physiological effects of fluometuron on some unicellular algae. *Weed Sci.*, **16**, 296.

ST. JOHN, J. B. (1971). Comparative effects of diuron and chlorpropham on ATP levels in *Chlorella*. *Weed Sci.*, **19**, 274.

TCHAN, Y. T. (1959). Study of soil algae. III. Bioassay of soil fertility by algae. *Pl. Soil*, **10**, 220.

VENKATARAMAN, G. S. & RAJYALAKSHMI, B. (1971). Tolerance of blue-green algae to pesticides. *Curr. Sci.*, **6**, 143.

WALSH, G. E. (1972). Effects of herbicides on photosynthesis and growth of marine unicellular algae. *Hyacinth Control Journal*, **10**, 45.

WRIGHT, S. J. L. (1971). Degradation of herbicides by soil micro-organisms. In *Microbial Aspects of Pollution*, (G. Sykes and F. A. Skinner, eds). London: Academic Press.

WRIGHT, S. J. L. (1972). The effect of some herbicides on the growth of *Chlorella pyrenoidosa*. *Chemosphere*, **1**, 11.

ZWEIG, G., HITT, J. E. & MCMAHON, R. (1968). Effect of certain quinones, diquat and diuron on *Chlorella pyrenoidosa* Chick (Emerson strain). *Weed Sci.*, **16**, 69.

Author Index

Numbers in italics are pages on which references are listed at the end of the paper.

Subject Index